How to Rebuild Your
BIG-BLOCK CHEVY

By Tom Wilson

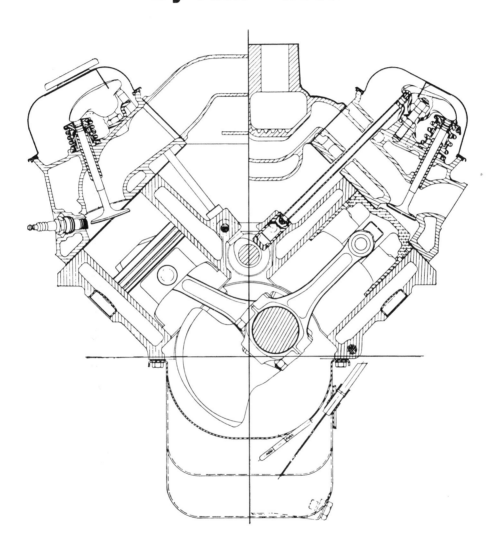

Photos: Tom Wilson; others noted

Published by HPBooks
A Division of Price Stern Sloan, Inc.
360 N. La Cienega Blvd., Los Angeles, CA 90048
ISBN 0-89586-175-5   Library of Congress Catalog Number: 83-81691
©1983 Price Stern Sloan, Inc.   Printed in U.S.A.

10 9 8 7 6 5

Contents

Introduction 3
1 Time To Rebuild? 4
2 Engine Removal 14
3 Parts Identification & Interchange 25
4 Teardown 49
5 Shortblock Reconditioning 64
6 Cylinder-Head Reconditioning 94
7 Engine Assembly 108
8 Engine Installation 136
9 Tuneup 157
Index 158

ACKNOWLEDGMENTS

Writing a complex book such as this requires the help of many people. With apologies to those whom I may have overlooked, I'd like to thank the following people for their generous support:

Starting close to home, friends stepped forward to help. Steve Beegle, Mike Breining, Neil Emory, Larry and Mike Granzella, Randy Lyles of British Masters Racing, Greg Pedersen and the gang at Fallbrook Foreign Car Center, Rick Shahan and George Winch were all there when it counted.

Dragmaster Machine Shop provided me with years of "rat" experience through George Lane, Dode Martin, Neil Morrison and Jim and Tom Nelson. Neil was especially helpful, staying after hours to machine yet another Chevy for the camera.

Harold Bridges and Monty Thill of Motor Machine & Supply gave professional advice.

More pro help came from Bob Bub of Cloyes Gear and Products, Gary Chilcher of Lakewood Industries, Terry Davis of TRW, Cal DeBruin of Sealed Power Corporation, Bill Dismuke of NHRA, Bob Lopez of Federal-Mogul Corporation, and Robert Morris of Fel-Pro Inc.

Chevrolet's help came from many directions. Joe Keith and Duane Poole assisted from their management positions. At Guarantee Chevrolet, Santa Ana, California, Bob Dean and Danny Wilheight endured endless questioning. Security Chevrolet, Vista, California, provided Mike Ponomarenko, Phil Stauch and Phil Vogel to answer questions.

There was more help with production: Phyllis Bryan typed unreadable copy, Bill Ahrend and Myron Hemley of Fallbrook Camera helped with photos, as did Marge Binks at Photic, Inc.

Special thanks to Jan for her help and patience.

ANOTHER FACT FILLED AUTOMOTIVE BOOK FROM HPBooks

Introduction

Mark IV Chevrolet has been installed in a wide spectrum of vehicles, from sports cars to trucks and everything in between. Big-block-powered 'Vette, beginning with this '65 425-HP 396, gave the best performance-per-dollar value ever. Photo courtesy of Chevrolet.

As modern engines shrink yearly, the big-block Chevy becomes more and more impressive. Chevrolet calls this engine the *Mark IV*. Its size, rugged construction and Herculean power have provided the motivation for everything from sedate passenger cars to "time-warping" race cars. The big-block Chevy was an instant hit at its 1965 introduction. Since then, it has powered medium- and light-duty trucks, and passenger cars such as: the Chevelle, Camaro and Corvette sports models and the full-size Impala. Each engine part was made for durability and power—qualities not always found in today's small, fuel-stingy engines.

Introduced with a 396 cubic-inch displacement (CID), bore and stroke measured 4.094 X 3.760 in. But the big news was valve arrangement. Instead of placing all the valves in a row as was done with its small-block predecessor, Chevy engineers canted both the intake and exhaust valves for improved breathing. Performance people lovingly refer to the big-block Chevy as the *porcupine* or *rat motor*. It was an instant winner.

In 1966 a 427-CID version was added. The displacement increase came from a larger bore, 4.251 in.

Otherwise the engine was unchanged. Aluminum heads and blocks were options—at extra cost. Most 427s were installed in Corvettes. The 427 was dropped in 1969.

Chevy got sneaky in 1970. The 396 engine was built with a 0.030-in. overbore, giving it an actual displacement of 402 cu in., but dubbed it a 400. Engine decals and car plaques continued to label the engine incorrectly as a 400 until 1972 when it was finally called a 402—the engine's final year of production.

In 1970, Chevy introduced a really big-displacement version of the big-block, the 454. The 427's 4.251-in. bore was used, but a new 4.00-in.-stroke crank was fitted to the new block casting. Of all Chevy passenger-car big-blocks, the 454 is the most different. But that isn't saying much; these engines are remarkably similar, so almost all parts interchange.

The 454 proved to be a versatile and long-lived engine. By 1976, ever-tightening emission and mileage demands ended its installation in passenger cars. Light and medium trucks were affected in '79-'82. The 454 was only available in trucks of 1/2-ton capacity or more in 1979. The

big-block wasn't available at all in 1980—'82. Due to popular *demand*—small-blocks couldn't handle the loads—the big-block became available again in 1983 in 1-ton and larger medium-duty-truck chassis. High-performance service parts and engines are still being manufactured because these engines are very popular for boat- and drag-racing use.

Mark IV big-blocks are also used—with few changes—in large trucks. These versions displace 366 and 427 cu in. Most parts will interchange with passenger-car engines.

Rebuilding a big-block Chevy may sound like an insurmountable task—not so. Any engine work becomes easier with information. This book provides the information you'll need to rebuild your big-block Chevy quickly, easily and professionally.

While writing this book I talked with many big-block Chevy experts, both factory-trained and garage-experienced. I rebuilt several engines for the photos as well as to gain first-hand experience. These engines are simple and fun to work on if you follow certain rules.

Read through the book and study the pictures before you start to work. You'll be better able to handle the upcoming jobs with the proper tools and planning. Don't skip steps because they seem unnecessary. If they were unnecessary I wouldn't have put them in the book.

Double-check your work at each step. Double-check the double-checks and be extra sure to check others' work. Machine-shop and parts men make mistakes too.

Catch the mistakes before they become serious. The time to discover the mismarked bearings is before you assemble the engine, not afterwards.

Finally, work safely. Keep your work area clean. Don't let children, pets or the curious get in your way. Make sure all lifting and support equipment is strong enough for the loads. If in doubt, get an experienced helper. The big-block Chevy is heavy—about 680 pounds. Knock off about 150 pounds for an all-aluminum ZL-1.

Enough said. Let's get to work.

1 Time To Rebuild?

Before deciding to rebuild your big-block Chevy you should first determine that a rebuild is necessary. A very thorough tuneup, including a carburetor rebuild and a distributor check, can solve a lot of problems. Big-block Chevys are strong, understressed engines with a reputation for long life. They don't live forever, but rebuilds aren't normally necessary until well over 100,000 miles. Oil and gas consumption will be excessive and power will be down if a rebuild is necessary.

Operating conditions greatly affect engine life. Example: A pickup truck driven offroad will inhale more dirt than a passenger car used for highway driving, resulting in higher-than-normal wear. Easy freeway miles don't cause engine wear like high-revving sporty driving. How a car was serviced counts too. Infrequent oil and filter changes can shorten an engine's lifespan considerably.

This is not a new engine. Photographed nearly 14 years after it was installed, 1969 Corvette with 435-HP 427 is kept in top condition through proper maintenance.

OIL CONSUMPTION

Engine oil consumption is a good way to judge its internal condition. Sealing between moving parts—particularly valve stems and guides, and piston rings and cylinders—is the major factor governing oil consumption. The more an engine wears, the more clearances increase. As clearances increase, oil consumption increases *and* oil pressure decreases.

Because some oil is burned in normal operation, determining what is abnormal depends on your particular engine. Certainly 1000 miles per quart is not bad. Between 1000 and 500 miles per quart indicates problems, but not crippling ones. Less than 500 miles per quart indicates a serious condition that should be fixed immediately.

Don't be misled by a leak. Puddling on the garage floor points to a faulty gasket or seal, not worn engine parts.

If oil consumption is high, check first for worn or broken piston rings, bad valve-stem seals, and worn valve guides. These problems don't cause leaks. They allow oil to enter the combustion chamber where it is burned.

Because oil doesn't burn completely, it shows up as blue smoke from the tailpipe. Best times to spot oil smoke are immediately upon starting the engine and when applying the accelerator after descending a hill or after coasting.

Oil pressure—or lack of it—is also a good indicator of engine condition. Two types of oil pumps are used on Chevy big-blocks; the standard unit and a high-volume model installed in some high-performance versions. The standard pump should manage 40 pounds-per-square-inch (psi) at 2000 rpm or more and 60 psi for the high-volume version.

If the oil-pressure gage indicates slightly less than these pressures while driving, but drops to about 10 psi at idle, your engine has excessive journal-to-bearing clearances. This assumes your engine has gone many miles, is filled with 30-weight or thicker oil and is thoroughly warmed up. If you don't have an oil-pressure gage, but the warning light flickers on at idle, there is a definite problem. The warning light shouldn't come on unless oil pressure drops well below 10 psi!

Piston Rings—Passenger-car big-block pistons use three rings. The top two primarily seal compression and combustion pressures. The bottom ring *controls* oil by spreading it evenly over the cylinder wall without allowing an excessive amount enter the combustion chamber. Some oil must pass the oil ring to lube the compression rings, piston and bore.

The 366- and 427-CID truck big-blocks have an additional compression ring between the second compression ring and the oil-control ring. This fourth ring helps the compression rings seal the combustion chamber.

Ring wear results from normal friction as each piston travels up and down its bore. Wear is greatly accelerated by grit entering the engine through the induction system. As ring wear increases, more oil enters the combustion chamber during the intake stroke.

Conversely, compression and combustion gasses enter the crankcase on the compression and power strokes, respectively. This is called *blowby*. Blowby gasses are recycled back to the intake manifold by the positive crankcase-ventilation (PCV) system.

You can make a quick check for blowby by watching a disconnected PCV valve or the dipstick or oil-filler tubes. Smoke from these is blowby.

Valve Guides—Worn guides and damaged valve-stem seals allow excessive oil past the valve stems and into the intake or exhaust ports. Because of the low relative pressure in an intake port, oil is drawn down between guide and valve stem into the combustion chamber. Exhaust ports are pressurized, so exhaust gasses blow out the top of exhaust-valve guides: more blowby in the crankcase.

When an engine with worn guides is shut off, oil drains down both intake and exhaust guides, causing that puff of blue smoke on startup. Another sign of of worn guides occurs following periods of high manifold vacuum—accelerating after idle or decelerating. Look for a puff of blue smoke at the tailpipe.

POOR PERFORMANCE

Both power and fuel consumption must be checked to determine if engine performance is suffering. If oil consumption is not excessive, but power is low and fuel consumption is high, give the engine a tuneup. Take the car to a reputable tuneup shop, preferably one with a chassis dynamometer. The dyno will provide you with an exact horsepower figure to work with.

Both fuel- and oil-consumption figures are easily obtained by keeping accurate records. But determining power output takes more than a seat-of-the-pants feel. Power loss is too gradual to be noticed, especially in a high-torque engine such as the big-block Chevy. Dyno results are much more accurate.

PERFORMANCE PROBLEMS

Piston-ring and cylinder-bore wear are suspect if power is down and oil consumption is high. If blowby is excessive, there will be a corresponding loss of compression and combustion pressures.

However, if—after a tuneup—the engine is down on power and mileage is poor, but oil consumption is not bad, check for: blown head gasket, burned exhaust valve, worn cam and lifters and carbon buildup.

Blown Head Gasket—Extremely low compression results from a blown head gasket. Accompanying the low compression is an immediate power loss. The lost pressure passes into the cooling system, an adjacent cylinder, to the atmosphere or a combination of these.

When compression and hot combustion gasses find their way into the cooling system, the coolant is heated, pressurized and forced out, eventually causing the engine to overheat. The engine will also warm up rapidly after a cold start. This type of combustion leak can be detected with a tester that draws air from the radiator and runs it through a small amount of chemically treated fluid. A change in fluid color indicates the presence of combustion gasses in the coolant.

High power losses occur when a head gasket leaks between cylinders. A blown head gasket that causes a leak between two cylinders greatly reduces the power output of the two cylinders. Power and mileage losses are then easily detected. This type of head-gasket leak also causes a rough idle.

A blown gasket that allows combustion-chamber pressures to vent to the atmosphere is easily detected. Similar to leaky exhaust manifold, you'll hear a loud ticking noise that occurs at 1/2 engine speed. As engine speed increases, so does the frequency of the noise.

Burned Exhaust Valve—Power loss due to a burned exhaust valve is significant. Because combustion pressure escapes past the exhaust valve and out the tailpipe, mileage and power will be poor. Later engines with more emission controls run hotter and tend to burn valves more frequently. It is not uncommon to find one or more burned exhaust valves in a '70-or-later big-block.

Worn Camshaft & Lifters—Lifters and camshaft lobes wear together. Because this wear is gradual you won't notice the power loss until after the lobe or lifter has worn through its hardened surface. Then wear accelerates and valve lift drops dramatically,

causing severe power loss. Cam wear is a frequent problem in later engines. It's not uncommon to find cams that need replacing in engines with barely 20,000 miles on the "clock," so don't overlook checking valve lift, page 12.

Camshaft and lifter wear is hard to detect in the early stages. There is no sudden power loss, nor problems such as missing, knocking, rough running or stalling. Cam-lobe and lifter wear reduces valve lift. This means that less air/fuel mixture is drawn into the combustion chamber during the intake stroke—with an intake-valve lobe or lifter—and less gas is forced out during the exhaust stroke. Result: a gradual loss of performance.

Carbon Deposits—One unfortunate side effect of babying an engine is carbon buildup. Low rpm coupled with light-load engine operation makes for relatively low combustion and cylinder-head temperatures. The resulting incomplete combustion creates excessive hydrocarbons in the exhaust. This helps the unburned carbon—a solid byproduct of combustion—attach itself to combustion-chamber, valve and piston surfaces. Higher engine speeds and loads, as in freeway driving, result in hotter, more-complete combustion. High-speed engine operation also reverses the carbon buildup by burning the deposits away.

Excess carbon in the combustion chamber has several harmful effects. First, it *reduces* the volume of the combustion chamber, *raising* the compression ratio. This is not harmful in itself, but with today's poor-quality gasoline, a higher compression ratio can cause the fuel charge to *explode,* or *detonate*—commonly referred to as *ping* or *knock*. This results in power loss and possibly severe engine damage. Damage can show up as damaged pistons, broken piston rings, deformed main-bearing caps or blown head gaskets.

Carbon deposits can also cause *preignition*—ignition of the fuel charge before the sparkplug fires. The fuel charge is ignited by red-hot carbon—much like the glow plug of a two-cycle model-airplane engine. Power loss also results, possibly followed by a burned piston. Preignition can also lead to detonation.

Carbon deposits also reduce power output by *shrouding* the valves, restricting flow to and from the combustion chamber. Shrouding occurs as carbon builds up on the backside of a valve head and around the valve on the cylinder head, either on the combustion-chamber or port-side of the valve. Although carbon shrouding reduces performance, it doesn't harm the engine.

What is dangerous is the chance of a small piece of carbon breaking loose and getting caught between the head of the valve and its seat. Then the valve cannot close fully, letting the burning air/fuel mixture burn the valve during the power stroke. Exhaust valves are more susceptible to this kind of damage than intakes.

That small piece of carbon could also get caught between the sparkplug electrodes. This shorts the plug, causing a *miss.* You'll have to clean the plug before it will fire.

The best cure for carbon deposits is prevention. Freeway driving burns away carbon; chugging around town "lays it on with a trowel." If you use your car mainly for short trips, take it out on the freeway for some occasional exercise. **Don't** pour carbon-dissolving "instant-overhaul" liquid in your engine. Sure it will loosen the carbon, but it will also break the carbon loose and cause the problems mentioned above.

Another way of preventing carbon deposits and controlling detonation is to install a quality water-injection system. The water not only prevents carbon buildup—existing carbon buildup will burn away. And a cooler fuel charge has less tendency to detonate.

DIAGNOSIS

Now that you know some of the things that can go wrong inside an engine, let's pinpoint the problems. Most tests can be done without special tools or knowledge, but a few are probably jobs for the pro. If you have taken good care of your big-block it might not have any serious problems, even though it may be time to rebuild. Now is the time to find out.

Noises—An engine noise can be hard to pinpoint—it may seem to jump around the engine compartment. You may become accustomed to a noise, you

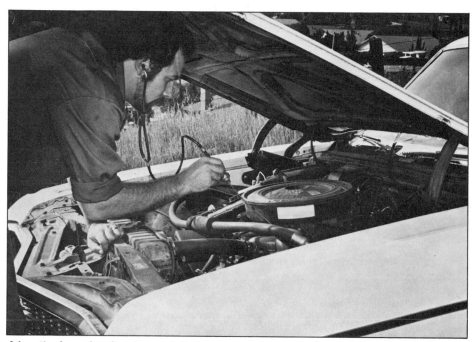

A length of wooden dowel or a stethoscope can be used to pinpoint internal engine noises. Stethoscopes work doubly well because they block-out distracting external noise while amplifying internal noise.

perhaps not even hear it, but your passenger keeps asking you about it. The problem is most mechanics do not, or cannot, differentiate between the many noises a car makes. You must learn to tune out those noises that give you no clues, such as those from the fan, exhaust or fan belt.

When you open the hood to listen, first eliminate the routine noises. Listen to the fan, then block that noise out of your mind. Then eliminate rattling sheet metal or other interferences. After you have the obvious sounds blocked out, the subtler internal engine noises should be more evident. The point is; try not to listen to the running engine as a whole, but to concentrate on each noise—one at a time.

Noises can be placed in one of two categories: those that occur with each crankshaft revolution and those that occur at every other crank revolution—or at engine speed and at half engine speed, respectively.

If you have trouble telling how often the noise occurs, hook up a timing light, then watch and listen. Noises occurring *twice* for each flash of the timing light are at engine speed. Most *bottom-end* noises occur with each turn of the crank. These are typically caused by bad rod or main bearings, pistons, rings or piston pins—wrist pins.

Noises occurring *once* for each flash of the timing light—at half engine speed—usually come from the valve train. Why? Because the camshaft turns at one-half crankshaft speed. Chances are the noise will be the ticking of a collapsed hydraulic valve lifter—*tappet.* Valve trains with mechanical lifters normally make some valve-train noise, even though adjusted properly. However one with too much clearance will clack similar to a bad hydraulic lifter.

One bottom-end noise occurs at half engine speed—*piston slap.* Why? Because piston slap occurs only on the power stroke. Continue reading for details on checking for this noise.

You may have seen a mechanic listening to an engine with a stethoscope. Besides assuring the proper "Doctor of Motors" look, a stethoscope is useful in pinpointing the source of a noise. A 3—4-ft piece of hose, wooden dowel rod or a screwdriver works almost as well. Use a rigid plastic hose if possible—it transmits sound better than rubber or any other soft material.

Unfortunately, sounds are also carried by the cylinder block, so you may have trouble pinpointing the noise. Whatever listening device you use—wood dowel, rigid plastic hose or screwdriver—hold one end against a solid part of the engine. Press the other end against your ear or touch it to your head behind your ear. Be careful, don't jab yourself.

Gaskets and air spaces damp noises. For instance, if you're listening for a loose timing chain, listen against a timing-cover bolt, not the cover itself. The cover is insulated by the gasket and the air space behind it. The cover bolt engages the block solidly, consequently it transmits noise better.

Main Bearings—Worn bearings create noise because excessive bearing-journal clearance causes reduced oil pressure. Result: The oil cushion between the bearings and crank journals is reduced. The crankshaft then moves and pounds main bearings with a deep, knocking noise. You'll hear it best standing several feet away from the car or sitting inside it. The sound may seem to stop when you open the hood because other noises can mask the low-frequency knock.

Test for worn main bearings by putting them under load. With an automatic transmission, put the car in **DRIVE** and apply the brakes. If the car has a manual transmission, feather the clutch in first gear. Apply the throttle, but don't let the car move. As the engine is loaded, a bad main bearing will knock louder. This is also noticeable when going up hills, accelerating or under any other condition where the engine is loaded.

Rod Bearings—Rod bearings knock for the same reason as mains—excessive clearance—and the noise is similar. But rod knocking is higher pitched than that from a main bearing, and it must be tested differently.

To test for rod knock, place the transmission in neutral and lightly rev the engine. Make sure you lift off the accelerator abruptly; engine rpm must peak sharply. When you do this, a bad rod bearing will rattle as it passes through the transition of being loaded, then quickly unloaded. Both

rod- and main-bearing tests should be done on a thoroughly warm engine—after at least a half-hour of running. The thinner the oil, the easier it is to hear a bad bearing.

Piston Noise—Unlike noise from excessive bearing clearance, piston noises are best detected when an engine is cold. This is because piston slap is caused by excessive piston-to-bore clearance at the piston skirt. Clearance—and noise—is greatest when an engine is cold and the piston hasn't expanded. When the cylinder containing the faulty piston fires, the excess clearance between the bore and piston lets the piston impact, or *slap,* against the *thrust-side* of the cylinder wall—the right side, as viewed from the flywheel-end of the engine.

A collapsed piston skirt makes a similar hollow, knocking sound. Disabling the cylinder will stop or greatly diminish the noise caused by either problem. Retarding the spark is another way to make this test. The problem is, it doesn't tell you which piston is at fault. Don't forget to reset ignition timing when you finish.

With the engine idling, disconnect the plug leads or ground each plug one at a time to disable the cylinders. **With an electronic ignition, don't disconnect the plug wires; use the grounding technique,** page 9. Otherwise you risk damaging an expensive ignition system due to excessive voltage buildup. When the noise ceases, you've found the "slap-happy" piston!

Hydraulic Lifters—All but the highest-output big-block Chevys are fitted with hydraulic lifters. A hydraulic lifter, or tappet, consists of a lifter body and plunger. The *lifter body* is the outer shell that moves up and down in its bore in the block, immediately above the camshaft. Its lower, closed end—*foot*—rides on a cam lobe. Within the lifter body, a piston—*plunger*—bears against a small amount of oil while its concave upper end seats the pushrod.

A small hole in the side of the lifter body allows oil from a cylinder-block oil gallery to pressurize the inner chamber of the lifter. The oil-pressurized lifter provides a "cushion" between pushrod and the cam lobe. More importantly, the lifter plunger is free to travel within limits in the lifter body, taking up *lash*—clearance—in the valve train.

As long as oil pressure is maintained in the lifter body and the plunger hasn't exceeded its travel limit, the valve train is kept in perfect adjustment with *zero clearance*. With valves, rockers arms, pushrods and lifters in constant contact, wear and noise are minimized.

"Gone solid" is a good description of what happens to a collapsed hydraulic lifter—it sounds similar to a solid lifter with excessive clearance. There are two major causes of collapsed lifters: a stuck check valve or a stuck plunger. When a plunger sticks, oil pressure can't overcome the resistance between the lifter body and plunger to take up valve-train clearance. When a check valve sticks, the lifter fills with oil, but it bypasses the valve and the oil leaks out the other side when loaded by the valve train.

At first, a bad hydraulic lifter makes a soft, ticking sound. When fully collapsed it develops a harsh clack. If a hydraulic lifter begins to make noise, check the oil; it should be up to the full mark and clean. Low or dirty oil is the number-one killer of hydraulic lifters. But a collapsed hydraulic lifter won't quiet down when you pour in a few quarts of oil. Fresh oil won't save ruined lifters—or the cam the lifters just pounded to death!

To find the bad lifter, remove the valve cover from the head on the side the noise appears to be coming from. Idle the engine. Using about a 0.020-in. feeler gage, slip the gage between each valve tip and rocker arm. When you get to the one with the bad lifter, the noise should change markedly.

If the lifter is just starting to tick, try this fix. Fill a squirt can with carb cleaner; remember it's a powerful solvent, so follow the instructions carefully. Remove the rocker cover and squirt a little cleaner down the pushrod to the lifter. A lifter that's gummy from old oil or low-speed driving will sometimes clear itself. If the carb cleaner doesn't work, replace the collapsed lifter.

A hydraulic lifter can also be noisy when it wears excessively. When new, a hydraulic-lifter foot is *convex*—rounded outward at the center. If this surface becomes so worn that it is *concave*—rounded inward at the center—wear may exceed the travel of the plunger. Ex-

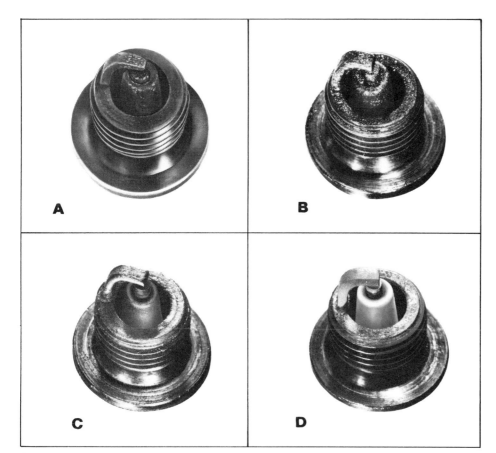

Sparkplugs can be thought of as removable portions of the combustion chambers that let you *read* what's going on inside. For instance, plug A is worn out. The electrodes have rounded edges and the insulator is pitted. Later, hot-running smog engines may give the center porcelain a greenish cast once the plug has been overheated. Replace worn-out plugs. B is oil-fouled, as indicated by the wet, shiny black coating. Worn rings, cylinder walls and valve guides and seals cause this condition. C is fouled by an overrich air/fuel mixture. Carburetion problems and slow-speed driving will foul plugs with these dry, flat-black deposits. D is a normal plug. The electrodes still have sharp edges and the insulator is not fouled, although it will have a slight tan color. This plug can be cleaned, gapped and reinstalled. Photos courtesy of Champion Spark Plug Company.

cessive clearance and noise results. This problem can be only be corrected by replacing the lifter.

Mechanical Lifters—A soft sewing-machine-like ticking is normal with mechanical lifters, but excessive noise means trouble. The quickest way to find the loose rocker arm is to use the feeler-gage method. Slip a 0.020-in. feeler gage between each rocker arm and valve-stem with the engine at idle. When you gage the loose rocker arm, the noise will diminish.

Normally, noisy mechanical lifters are due to incorrect adjustment. The cause may be a bent pushrod or a loose rocker-arm adjustment nut. Check for both while the valve covers are off.

Adjust intake valves to 0.024 in., exhausts to 0.028 in.; 0.022 in. and 0.024 in. for 430 HP 427s. Valve adjustments should be made when the

engine is at operating temperature and running. For other than a stock mechanical cam, check manufacturer's specifications.

Heat-Riser Valve—A noise may sound like an engine's death rattle, but turn out to be relatively minor. Example: A stuck heat-riser valve can be very noisy. Located on the left exhaust manifold, the heat-riser valve is controlled by a *bimetal* spring—such as that found in a thermostat. With a cold engine, the spring shuts the valve, forcing exhaust gasses through a passage in the intake manifold. This warms the incoming air/fuel mixture for smoother startup and gas-saving, cold-weather performance.

A broken bimetal allows the valve to be whipped back and forth by pulsating exhaust gasses. It sounds similar to a broken valve spring. Temporarily wire the valve counterweight to the exhaust pipe to stop the noise.

If you want your engine to operate smoothly when cold, you'll need to replace the heat riser with one that works. Later engines use a vacuum-controlled heat-riser valve that doesn't have the spring-breakage problem.

Fuel Pump—Before you tear the engine down because of a bottom-end noise, check the fuel pump. Fuel-pump noise will normally be a fairly loud, double click at *half engine speed*. It's often mistaken for a bad rod.

If you want to be sure the fuel pump is at fault, disconnect the fuel lines, plug them and remove the pump. Run the engine—there'll be enough fuel in the carburetor for a couple of minutes of idling. If the noise ceases, chances are the fuel pump is bad.

SPARKPLUG READING
You can get a general idea of your engine's condition by *reading* or examining, the sparkplugs. A plug gives a clue as to the condition of its cylinder, so keep them in order. Remove each plug-wire boot with a pulling twist. Don't yank on the wire. Plug wires are fragile, so avoid bending and pulling.

Whether it's excess-oil leakage past the piston or intake-valve guide, the sparkplug from a well-oiled combustion chamber will be coated with a wet, shiny black deposit. If the plug is shiny black, rub it into your palm. The black residue will not wipe away. If the leak isn't severe, the plug will be coated with a dry, washed-out tan crust. Such dry deposits are usually found in engines with minor oil-burning problems and worn sparkplugs.

Dry, black soot indicates a rich air/fuel mixture. This sooty black deposit is not shiny and will practically disappear when rubbed. A rich fuel charge is the result of a dirty air filter, or an improperly adjusted or malfunctioning carburetor. A cylinder with low compression due to poor sealing at the cylinder or a valve won't burn its fuel charge completely and will *appear* to run rich.

The opposite extreme is a lean air/fuel mixture. A lean mixture burns too hot, frequently resulting in burned exhaust valves. At the extreme, it can also mean burned pistons. The plugs become coated with a white residue. This is common

Vacuum gage measures an engine's total pumping capacity. Although it will give a quick indication of the condition of an engine, further testing is required to pinpoint a problem.

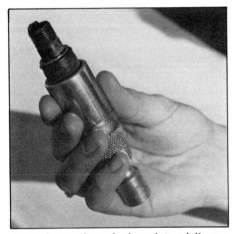

Close inspection of plug showed it was more than a little fouled. Large, solid deposits indicate oil burning. When fouling is this bad there is definitely something wrong with either the rings or valve guides.

with '71-and-later "emission-control" engines that are *calibrated* to run lean. This was done in an effort to meet the stiff mileage and emissions regulations mandated by the Federal Government. A slightly richer mixture or colder heat-range plug should cure the problem.

Later engines with emission controls have a habit of using up sparkplugs at an incredible rate. Sometimes 5000 miles is all you'll get from a set before they "burn up." With the big-block Chevy this problem is unfortunately all too common. You can recognize an overheated plug by the eroded electrode and bubbled porcelain;

often with no trace of white residue. The sparkplug porcelain on many emission-control-equipped engines will have a faint greenish or tan color, indicating overheating and need for replacement.

Don't rely on plug readings for a perfect diagnosis. Many factors affect the color of plugs and there are better ways of determining cylinder condition. Rely on plug reading only as a quick check. From a rebuild standpoint, you are most concerned with excessive oil consumption, so watch for it.

TESTS

Cranking Vacuum Test—An engine acts as a suction pump to draw the air/fuel mixture into its cylinders prior to combustion. Other variables notwithstanding, the more air/fuel an engine draws in, the more power it produces. You can measure your big-block's suction ability with a vacuum gage. Besides the gage, you'll need a notepad and a friend to operate the starter—unless you have a remote starter.

Warm the engine, then connect the vacuum gage *directly* to the intake manifold. Some vacuum connections at the carburetor are *ported,* producing less vacuum than at the manifold. The power-brake connection at the manifold is a vacuum source.

Disconnect and ground the coil *high-tension* lead—the heavily insulat-

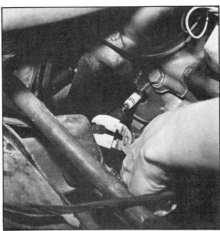

Check for a strong spark from a points-type ignition by pulling the lead off with insulated pliers. Hold the plug lead about 1/4 in. from the plug with the engine running. You can hear the spark as it jumps the gap. A good spark will make a sharp crack; a so-so spark is much more subdued. Don't do this with an HEI ignition or you'll blow the module.

DISABLING IGNITION

Two types of ignition systems are used on big-block Chevys. The first is the familiar point-type *window* distributor. This system was standard until 1974. The second ignition system is the HEI (High Energy Ignition) electronic system, offered as an option in 1974 and made standard in all GM lines in 1975.

The earlier point-type system is easily "turned off" by disconnecting the high-tension—heavily insulated—wire at the center of the distributor and grounding it to the intake manifold.

The HEI system has a small bundle of wires entering the distributor cap on the driver's side of the engine. This bundle connects the distributor to the electronic ignition module—the "brain." The wires enter the distributor cap in a quick-disconnect plug. By disconnecting this plug when you want to crank the engine, but don't want it to start, you'll keep from harming the ignition system. This is important to remember when performing a compression test, cranking for oil pressure or for any other reason. Use the jumper method when performing a power-balance test, page 10.

ed one. With HEI systems, remove the quick-disconnect plug at the distributor cap. Disabling or grounding an HEI ignition is vital to avoid extreme voltage build up. So ground or disable the ignition system. Otherwise

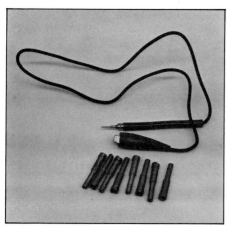

Using this type of kit is the simplest and safest way to check cylinder power balance, especially with an HEI ignition. Jumpers are installed between each plug and its lead so individual cylinders can be shorted without risking ignition-system damage. Photo by Tom Monroe.

you risk damaging an expensive ignition system or getting a voltage discharge into your body—a "shock" in either case.

Now, crank the engine while you watch the gage. It will take a few crankshaft revolutions for the needle to stabilize; then it should give a steady reading. If the needle continues to fluctuate, there is at least one poor-sealing cylinder.

Many conditions can cause this problem: worn cam lobe, collapsed lifter, incorrectly adjusted valve, incorrect valve timing, poorly sealing valves or rings or a leaking head gasket. Remember, cranking rpm must be reasonably high and steady. If the vacuum test shows a problem, a power-balance, compression or leak-down test will pinpoint the bad cylinder/s. High-performance cams normally give a low, bouncy vacuum reading; keep this in mind if your engine has a high-lift cam.

Power-Balance Test—This test shows how much each cylinder is contributing to an engine's power output. Hook up your engine to a sensitive tachometer such as that found on a tach/dwell meter. Warm up the engine, turn it off and disconnect all the plug wires at their plugs. Replace the plug wires in loose contact with the sparkplugs so you can easily remove each wire during the test. With an HEI ignition, use a jumper spring at each plug like those pictured.

Restart the engine and set idle to 900—1000 rpm so you can use the sensitive scale on your tachometer.

Note engine rpm. Remove the first plug wire with a pair of insulated pliers and ground it to the block, head or exhaust manifold. Or simply ground it if you are using jumper wires. After engine speed stabilizes, record the lower rpm on your pad. Reconnect or unground that sparkplug and go to the next one. After you've done all eight, shut off the engine and calculate rpm loss for each cylinder.

The greater the rpm loss, the more a cylinder contributes to total power output. The less the rpm loss, the less it contributes. It isn't so important how much rpm is lost as each cylinder is disabled, but that the rpm losses are nearly equal. All cylinders should fall within 25 rpm of each other. A differential of 40 rpm or more indicates a problem. Really bad cylinders won't show much, if any, of a drop. If the power-balance test reveals a potential problem, retest to be sure.

If your vehicle is equipped with a catalytic converter, don't let the engine run more than necessary when doing the power-balance test. Raw fuel from the disabled cylinder will cause the converter to overheat.

Compression Test—Just as a vacuum test measures an engine's ability to draw in air, a compression test measures its ability to compress that air. You can detect a poor-sealing piston, valve or head-gasket problem with this test.

First, warm the engine. Shut it off and remove all the plugs. Disable the ignition at the coil lead or module clip. Fit the compression gage to the number-1 cylinder. Some gages screw into the sparkplug hole, others have a rubber snout and must be held in place. Crank the engine with the throttle and choke plates fully open. As the engine cranks, you'll get a reading on each compression stroke. Count them and give each cylinder an equal number of strokes. It usually takes eight strokes to get a usable reading—sometimes more. Write down the results. Test all eight cylinders.

Again, as in the power-balance test, absolute numbers don't mean as much as how the cylinders compare. In this case, it's the pressure *differences* between the cylinders. Tune-up specifications call for 160 psi for most big-blocks. However, you may get readings from 70 to 200 psi. Cranking speed, the particular com-

pression tester, camshaft *profile* and other factors influence readings. To evaluate compression readings, use the 75-percent rule: The lowest cylinder should be within 75% of the highest. For example, if the highest pressure reading is 165 psi, the lowest acceptable figure can be found by multiplying 165 by 0.75, or 165 X 0.75 = 124. The lowest-reading cylinder must show at least 124 psi to pass.

If your big-block is one of the high-performance models, or you have installed a high-output cam, compression readings may be lower than average. This is a result of the increased *overlap*—the period when both intake and exhaust valves are open—built into most high-performance cams. So, if your 12.5:1 L-88 Corvette only pumps an even 100 psi, there's nothing to worry about. Altitude also affects compression. The higher you are, the lower the readings. This is simply because the engine is drawing air that is at a lower atmospheric pressure—start at a lower pressure; end up at a lower pressure.

If a compression reading falls below 75% of the highest reading, do a *wet test*. Squirt a teaspoonful of oil into the cylinder/s. Using a squirt can, insert its nozzle into the sparkplug hole and squirt the oil on the far side of the cylinder wall. Crank the engine for 15 seconds without the gage in place to spread the oil. Retest the cylinder. If the reading comes up noticeably—20 psi or more—poor sealing in the bore is the problem. If the readings *do not* increase, a valve, probably the exhaust, isn't sealing. Or the head gasket could be blown.

Another possibility is a cylinder that gives a higher-than-normal pressure reading. Strangely enough the oil ring could be at fault. If it is extremely bad, excessive oil will get to the compression rings and overseal that cylinder. You then get an excessively high, but *false* reading. Check the sparkplug for signs of oil. If oil is indicated, worn rings are a sure bet.

Leak-Down Test—A better way of determining cylinder condition is by performing a *leak-down test*. This test is superior to the compression test because some variables that affect a compression test are eliminated. These are: cranking speed, camshaft profile and wear, and compression ratio.

Leak-down test is better than a standard compression test because many influencing factors are eliminated. The engine is not turning, so cranking speed, mechanical compression ratio, valve timing, cam condition and internal oiling cannot affect the readings. Professional shops can perform a leak-down test.

Checking installed and open spring height with depth-gage end of vernier caliper. This will detect a worn cam lobe or stuck valve.

A leak-down test, unlike compression testing, can be compared directly; engine-to-engine and tester-to-tester. Another plus with leak-down testing is the ease of pinpointing the trouble spot. The bad news is price. Don't buy a leak-down tester unless you do a lot of engine diagnostic work—and own an air compressor. A tuneup shop should have one and will analyze your engine for a reasonable fee.

A leak-down tester doesn't rely on engine cranking to pressurize the cylinder being checked. Instead, it pressurizes the cylinder with air from an outside source—an air compressor. How much pressure depends on the calibration of the tester. After the engine is warmed, a fitting on the tester threads into the sparkplug hole of the cylinder being tested. With the cylinder at TDC on the compression stroke, it is pressurized by the compressed-air source through the leak-down tester. A gage on the tester indicates *percent leakage,* or how much additional air is required to maintain cylinder pressure.

Caution: When leak-down testing, stay clear of the fan and the accessory-drive belts. If the piston is not *on* TDC, the pressurized cylinder will cause the crank to spin 180° suddenly.

You can't use the starter to put the piston exactly where you want it. You'll have to do it by hand. Start with cylinder 1. With the plugs out and the rocker covers off, turn the crank clockwise by hand and watch the damper. Turn the crankshaft with a 3/4-in. socket and breaker bar on the damper bolt. When TDC at the damper is indicated as number-1 intake valve closes, the piston is at TDC. Mark damper in 90° increments for setting the remaining cylinders. After testing cylinder 1, turn the crank 90° and test cylinder 8. Follow the firing order: 1-8-4-3-6-5-7-2.

With a manual transmission you can put the transmission in gear and apply the parking brake to help keep the engine from turning over. You may even have to help with the breaker bar or the foot brake if your engine is in good shape.

It's another story with an automatic transmission. Someone to help, a long breaker bar and lots of muscle applied to the damper bolt is the answer with an automatic. But holding on gets tiring while testing all eight cylinders.

As air leaks from a cylinder, it can easily be traced by the noise it makes. If the intake valve is leaking, you'll be able to hear a hissing sound at the carburetor. The air cleaner must be off and the choke and throttle plates must be open. A leaky exhaust valve can be heard at the tailpipe. Leakage past the piston can be heard at the dipstick tube. A blown head gasket that leaks into the cooling system will cause bubbles to rise in the radiator *if the engine is warm*—the thermostat must be open. Leakage between adja-cent cylinders causes both to show poor readings, just as in the compression test. This leakage can be heard at the adjacent sparkplug hole.

What is an acceptable leak-down rate? Less than 10% leakage is excellent; 10—20% is not good; over 20% means trouble. Very bad cylinders can leak as high as 60%—100% if there's a *holed* piston, a blown head gasket or a burnt valve.

You may have trouble spotting a bad exhaust valve because the muffler will quiet the hissing. To double-check a cylinder with a high leak rate, get a CO (Carbon Monoxide) sniffer. Disconnect the leak-down tester and insert the CO-meter probe in the tailpipe. Squirt some carb cleaner in the cylinder and reconnect the leak-down tester while watching the CO meter. It will take a couple of seconds, but if the exhaust valve is leaking, the CO meter will give a full-scale reading in no uncertain terms.

Combustion-Leak Sniffer—You may have found what you think to be is a combustion leak into the cooling system through compression or leak-down testing. To confirm this, have a shop—probably the same fellow with the leak-down tester—*sniff* the cooling system with his color-changing-fluid tester. If combustion gasses are getting into the coolant, this test will confirm it.

CHECKING VALVE LIFT

If the engine has solid lifters, it's a simple matter to measure valve lift using the following technique. If the engine has hydraulic lifters, skip this section and measure the camshaft-lobe lift directly. The pressure created by the valve springs will cause a hydraulic lifter to bleed down—particularly on a high-mileage engine. This makes it difficult to get an accurate valve-lift reading. With solid lifters, make sure valve clearance, or *lash,* is set correctly.

If you have a dial indicator, use it. If you don't have a dial indicator, use the depth-gage end of a vernier caliper. If you don't have a vernier caliper, use a 6-in. machinist's scale.

To measure valve lift you must start with each lifter on the base of its cam lobe—piston at TDC on its compression stroke. Record the distance from the head to the top of the retainer with the valve closed. Turn the crankshaft until the valve is fully open—the spring is most compressed. The *difference* between the two measurements is valve lift.

If you are using a dial indicator, zero the indicator with the valve fully open. The indicator plunger must be square to the spring retainer and parallel to the valve stem. Set up the indicator so there is about 3/4-in. of plunger travel left. Zero the indicator. Crank the engine until the valve is fully closed. Read valve lift directly.

Compare readings after you've finished the last cylinder. Compare these readings to the chart, page 13. Recheck valves that don't open to specification. To get approximate lobe lift, divide valve lift by the rocker-arm ratio: 1.70.

Broken Valve Spring—While you have the valve covers off, check for broken valve springs. Early high-performance big-blocks—396s and 427s—had an Achilles' heal: fragile valve springs. Rather than using two springs to get the needed load, a single large-diameter-wire spring with a flat-wire damper was used. These high-rate, highly stressed springs tend to break near one end or the other. A partial coil is left to dance among the rocker arms and springs.

In order of increasing severity, several undesirable things can result: valve float at lower engine speeds, increased valve-stem and -guide wear, or a bent valve stem or pushrod. The

Extreme filth inside an engine usually means problems. Most evident is excess clearance at rocker tip and short installed spring height, indicating a stuck valve. One look at valve head proved this out, right.

Valve lift is best measured with a dial indicator. It is easier and the most accurate. Additional accuracy is obtained by removing the rocker arm and measuring lobe lift directly at the end of the pushrod. If you find a bad lobe, install a new cam.

valve floats because the absence of the coil gives less spring load. Stem wear is caused by side loading from the broken spring. And, if the wayward coil lodges between the coils of an adjacent valve spring, the spring goes *solid, or binds*—all coils touch—before the valve opens fully. A bent valve stem or pushrod results.

CAMSHAFT-LOBE WEAR

If a valve is not opening to specification, a hydraulic lifter is probably the culprit. To check the camshaft directly, remove the rocker arms. Measure lobe lift at the upper end of each pushrod. This gives the most-accurate measurement.

Set up the dial indicator so its plunger is in line with the pushrod and its tip is in the center of the pushrod. Rotate the crank so the lifter is on the base circle of the cam lobe—the indicator will be at its lowest reading. Zero the indicator by turning the dial face.

Rotate the crankshaft and watch for the highest indicator reading: This is lobe lift. Compare lobe-lift readings to those specified in the chart, page 13.

All lobes should be within 0.005 in.

If any of the lobes don't measure up to specification, but the engine has good compression and oil consumption is OK, you can replace the camshaft *and lifters*. Never mix old lifters with a new camshaft. Old lifters will quickly destroy the lobes of a new cam.

CONCLUSIONS

Let's review the results of your tests and put them in perspective. If your engine shows any signs of bottom-end wear, now is the time to rebuild. Low oil pressure, metal particles in the oil, knocks—all are danger signs. Problems such as these don't go away. And the longer you wait to fix the problem, the worse it will get and the greater the potential cost to do the rebuild.

If the bottom end appears to be in good condition, but cylinder sealing is not good, it is best to rebuild soon. However, if power is down, but the cylinders and rings are marginal, and the bottom end is OK—oil pressure is still high—you can squeeze a few more miles out of your engine before rebuilding.

A common situation is where all appears to be OK, with the exception of oil consumption due to valve-guide or -stem wear. A smart mechanic does a complete rebuild under these conditions, rather than "making do" with only a valve job. Probability is that oil consumption won't change much, even though oil loss down the guides has been controlled. This is caused by lower and higher combustion-chamber pressures due to improved sealing at the valves. Excess oil will come up past the pistons into the combustion chambers and increased blowby will pressurize the crankcase, forcing oil into the PCV system.

If the engine is just down on power, and all tests point to only a worn cam or collapsed lifters, your best bet is to replace the cam, lifters and timing chain. Remember, cams and lifters must be changed as a complete set. Replacing just the cam and not the lifters will ruin both when you first run the engine.

If you're in doubt, a rebuild is the best. Repairs will only get more expensive if you continue to drive the car.

		INTAKE		EXHAUST	
CID	**Year/HP**	**At Lifter**	**At Valve**	**At Lifter**	**At Valve**
366 Truck	all	0.234	0.398	0.234	0.398
427 Truck	all	0.234	0.398	0.253	0.430
396	1965/ 325	0.234	0.398	0.234	0.398
	425	0.306	0.500	0.306	0.500
	1966	0.271	0.461	0.282	0.480
	1967/ 325, 350	0.271	0.461	0.271	0.461
	1968– 1969/ 265, 325	0.234	0.398	0.234	0.398
	350	0.271	0.461	0.271	0.480
	375	0.306	0.520	0.306	0.520
402	1970– 1972	0.234	0.398	0.234	0.398
427	1966/ 390	0.271	0.461	0.282	0.480
	425	0.306	0.520	0.306	0.520
	1967/ 385, 390, 400	0.271	0.461	0.271	0.461
	425, 435	0.329	0.559	0.341	0.580
	1968– 1969/ 385, 390, 400	0.271	0.461	0.271	0.480
	425, 430	0.306	0.520	0.306	0.520
	1968/ 435	0.329	0.559	0.341	0.580
	1969/ 435	0.306	0.520	0.306	0.520
454	1970/ 345	0.234	0.398	0.253	0.430
	360	0.266	0.451	0.282	0.480
	390	0.271	0.461	0.282	0.480
	450	0.306	0.520	0.306	0.520
	460	0.306	0.520	0.323	0.550
	1971/ 365	0.271	0.461	0.282	0.480
	425	0.306	0.520	0.306	0.520
	1972	0.271	0.461	0.282	0.480
	1973– 1976	0.259	0.440	0.259	0.440
454	1975– 1979	0.234	0.398	0.234	0.398
Light Truck	1980–	0.234	0.398	0.253	0.430

2 Engine Removal

All-aluminum, ram-air ZL-1 427 is the epitomy of high-performance production engines. Cast-iron exhaust manifolds were used rather than aftermarket tube headers shown. Photo courtesy of Chevrolet.

Although not as involved as the rebuild itself, engine removal and installation are big steps in the engine-rebuilding process. If removal isn't done right, it can bring on tremendous headaches during installation. And both pose the most danger if approached haphazardly. So exercise care and planning so you'll save time and trouble, and avoid potential injury during engine removal *and* installation.

PREPARATION

Engine removal requires some special tools. The car or truck must be raised and supported, fluids drained, the engine lifted, and the fenders protected.

You probably have some of these tools: drain pan, jack stands, creeper and fender protectors. Two pieces of equipment you'll probably have to rent are: a *cherry picker* and floor jack. A cherry picker, or engine hoist, is a mini-crane mounted on casters. Its arm—powered by a hydraulic jack—extends over the engine compartment to lift the engine. The

cherry picker is the safest, easiest way to remove an engine.

The big advantage of a cherry picker is that once the engine is raised high enough to clear the body work, you can move it away from the car. The car can stay put. Other methods, such as the time-honored shade tree and chain-hoist, are usually inadequate for an engine as heavy as the big-block Chevy. And if the engine can be lifted safely, the car must be rolled out of the way before the engine can be lowered to the ground. Regardless of what you choose to use, remember that when you are selecting a cherry picker, or other lifting device, the engine you're lifting weighs about 700 lb. Get the *biggest* cherry picker you can find, even if it means paying more—$20 a day is reasonable.

Also think big when shopping for a floor jack. Big-blocks are installed in big vehicles. You'll need no less than a 1-1/2-ton-capacity floor jack with a beefy frame that won't flex under side loads.

Clean Engine—Removing the engine

will be quicker and easier with a clean engine and engine compartment. The job will also be safer and a lot more enjoyable. Spend a couple of quarters using the engine degreaser at a car wash. Blast off the normal road grime. Heavy grease and mud will require a putty knife and a can of engine degreaser. Steam cleaning is better if the engine is really caked with crud. Before you spray down the engine compartment, use plastic bags to cover the distributor, alternator, other electric components and the carburetor.

Where To Do It?—Before you decide where to pull the engine, plan ahead. Remember, once the engine is out, the car will be difficult to move. And the engine must be moved to where it will be torn down. You won't have much of a problem if you can do the removal on your own driveway. But if you live in an apartment, or where working on cars is prohibited, you've got a problem. Remember that the car is going to be immobile for no less than two weeks. If you have trouble finding a spot, look in the phone book

Under the wrench is something you'll need throughout the rebuild process—rags. Consider opening an account with a rag service. Or, if you have a friend already getting rags, ask him to increase his allotment to cover your needs. Of course, there are always old T-shirts and socks.

for a *do-it-yourself* workshop, or ask for space behind a nearby garage.

Try to do the engine-removal job on a concrete or asphalt surface. It can be done on dirt, but then you can't use a creeper. A large sheet of cardboard works best on dirt. And a cherry picker is very difficult to move on dirt. So, if you have to work on dirt make sure it's well packed. Another consideration: Dirt is a poor foundation for jackstands, so be extra careful. Put a 1-ft-square piece of 3/4-in.-thick plywood under each jack stand to keep it from sinking. Do this on asphalt too. A jack stand will sink in hot asphalt like it's mud.

When you have your site staked out, a floor jack and a hoist ready—along with drain pans, jack stands, hand tools and a couple of arm-twisted friends—you're ready to start.

ENGINE REMOVAL
Fender Cover—Place fender protectors—old blankets or throw rugs will do—on both fenders before starting. They will protect the fenders from buttons, belt buckles and tools. Leaning over the fenders become more comfortable, too. And tools won't slide off the fenders as easily.

Battery—Unbolt the ground, or negative cable (−) first, then the positive (+) at the battery. If your car has an automatic transmission, leave the battery in place. Use the starter motor to

BE ORGANIZED
Every mechanic who ever removed an engine has felt the urge to open the hood and tear into the job, yanking off hoses and wires helter-skelter until the engine is sitting on the garage floor. Mechanics who have submitted to this urge have also felt great frustration during installation when they couldn't remember where all the hoses and wires went.

An unaided memory just can't recall where everything goes—so don't rely on yours. If you have five hoses to connect, there are 120 ways in which they can be connected—and 119 are wrong! Don't guess, use masking tape and a felt-tip marker on every hose and wire you disconnect. Sometimes marks wear off, especially those made with ball-point, so write firmly. A label maker is excellent for marking—use one if you have one.

To store nuts and bolts, use old coffee or soup cans. Mark them on the outside so you can find what you want quickly. The more specific your categories the better—only three bolts in a bag is a lot better than 30 different bolts mixed together.

A camera can be a great help too. A few snapshots before things are disconnected can be invaluable. You'll then have the "original" engine for reference during installation. This advice is especially important if you own a late-'70s big-block covered with miles of vacuum hoses.

This book is as complete as possible but you will not find a hose-routing guide—there are just too many combinations to list. So, mark all of the connections clearly or pay for it later in wasted time and an engine that may not run right.

Heavier, baked-on crud is best handled with a steam cleaner. Look in the phone book for a gas station that performs this service. For about $20 you can stand by and watch *all* the dirt float away.

It takes two to remove a hood. First, mark around bolts or hinges so you can realign hood quickly when installing it. After marking, remove the hood bolts—start with the front ones. Keep a hand under the rear corner of the hood to prevent it from slipping back and scratching the fenders or cowl. Rags taped to hood corners help, too.

turn the crank to gain access to the converter bolts. With a manual transmission, remove the battery and clean its surface and terminals with baking soda in some warm water, a water hose and a wire brush. Store it safely, away from your work area. Remove the coil high-tension lead at the distributor or unplug the module lead on an HEI distributor.

Drain Oil—With a wrench for the oil-pan drain plug and drain pan, drain the crankcase oil. When the oil is completely drained, replace the drain plug so it won't get lost. Remove and discard the oil filter. Early engines have canister-type filters; later engines use spin-ons. To remove the canister type, remove the bolt from the end of the canister. Simply unscrew a spin-on filter. In either case, place the drain pan underneath so oil will drain into it as the filter is removed.

If you can't budge the spin-on-type filter with a standard strap-type filter wrench, drive a large screwdriver through the filter. Make sure the drain pan is underneath before you pierce the filter. Position the screwdriver so it won't interfere with turning the filter. The screwdriver will allow you to apply the needed leverage to loosen the filter. Let the oil drain for some time so it ends up in the drain pan, not on the driveway or garage floor.

Remove Hood—Before you unbolt the hood from its hinges, trace or scribe a line around the corners of

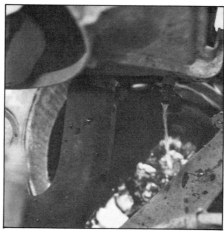

Open radiator petcock to drain radiator. Coolant will drain quicker if radiator cap is removed. If the coolant is fresh, save it by draining into and storing it in *clean* containers.

each hinge on the hood inner panel—don't go through the paint if you use a scribe. These reference marks will help you line up the hood correctly when it's replaced.

An alternate method is to drill an 1/8-in. hole through the hinge and hood inner panel. Be careful! *Don't drill up through the hood.* When it comes time to install the hood, an awl or small center punch inserted through the holes in each hinge and hood inner panel will align the hood exactly.

To protect the fenders, tape a folded shop rag or some cardboard to each rear hood corner. Then, with a friend at one side and you at the

Remove radiator hoses by first loosening the clamps. Don't pry off hoses. If you can't break them loose by twisting, slit ends of hoses with a knife and peel them off.

other, remove the bolts. Start with the front bolts; they're easy. You'll need three hands as you remove the rear bolts. When the hood comes free you'll be surprised at how heavy and awkward it is—definitely not a job for one person.

If you're working with a Corvette, a third friend is needed to hold the hood upright after the hood prop is unbolted. He can stand at the front of the car to support the hood while you and your other friend remove the hinge-to-hood bolts.

Place the hood upright in an out-of-the-way corner of the garage; protect the paint with a thick pad; an old mattress pad is perfect. To keep the

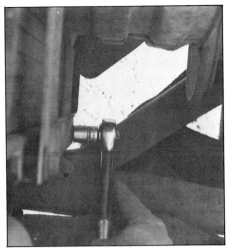

Unbolt fan shroud from radiator and lay it back over front of engine.

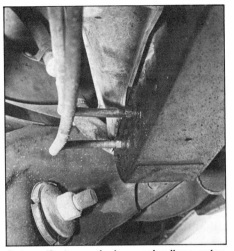

Automatic-transmission-cooler lines enter and exit cooler in bottom radiator tank. Use a flare-nut wrench to disconnect lines as shown. Use an open-end wrench and you'll round off the flare nuts. Once free, connect the lines with a length of clean fuel-line hose to keep ATF from siphoning from the transmission.

If your car has a cross-flow radiator, automatic-transmission-cooler lines will be on the passenger-side tank.

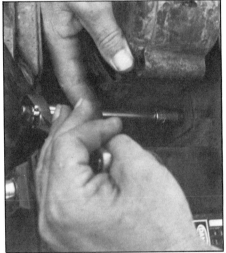

Remove the four bolts that hold radiator to radiator support. Carefully lift out radiator. Take care not to damage the fragile cooling fins.

With radiator out of the way, move fan shroud forward and out of engine compartment.

hood in place, wire the latch to a nail in the garage wall or stud. Another storage spot is on the car roof with a pad or large piece of cardboard in between, but only if the car is kept inside. Thread the hinge bolts back into the hood so they won't get lost.

Radiator Removal—Next on the drain list is the cooling system. With the drain pan underneath, open the radiator petcock. If you want to save the coolant, use a clean drain pan and store it in plastic jugs—*away from children*. Coolant will drain faster with the radiator cap off. Most of the coolant will drain, but some will remain in the bottom radiator hose and block.

The fan shroud is next; four bolts hold it to the radiator. Remove these and lay the shroud back over the fan. Remove both radiator hoses. Scrap them if they are more than two-years old, or if they are cracked or soft. Take care that you don't damage the radiator when removing the hoses. Don't pry them off. If you can't loosen each hose by twisting it, split the hose with a knife and peel it off.

Cars with automatic transmissions have two cooler lines running from the transmission to a bottom or side radiator tank. Undo the lines from the tank with a *flare-nut* wrench on the flare nut. A flare-nut wrench is a tool similar to a box wrench with a short section of one flat missing. This

allows the wrench to be slipped over the line and onto the nut, but retains its full-engagement capability. Use one to prevent rounding nuts.

After both cooler lines are disconnected, join their ends with a length of 5/16-in. fuel hose to keep automatic-transmission fluid (ATF) from siphoning from the transmission. Wire the lines up out of the way so they won't get damaged.

Remove the radiator. It is either bolted solidly in place or is set in rubber saddles and held in place with a clamp or clamps at the top. Remove the bolts or clamps and gently guide the radiator out of the chassis. Be extra careful not to bend the radiator

After removing shroud, unbolt a viscous-drive fan *from the water-pump flange.* If the pulley slips as you attempt to break the bolts loose, push down on drive belt to keep pulley from turning.

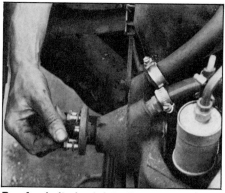

Run fan bolts back into the pump to keep them from being misplaced. This trick sure speeds up the reassembly process.

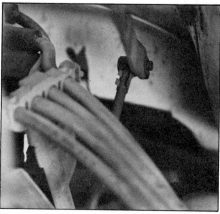

On large trucks, accelerator linkage is disconnected at the rod, over the bellhousing on the driver's side. Throttle rod is secured with a small clip to bellcrank arm extending from firewall. Snap off clip, slide rod out of arm and replace clip back on rod for safekeeping.

Passenger cars and pickups use either cable- or rod-type linkages. Cable type disconnects at carburetor by removing small clip. Disconnect rod type at firewall-end of rod.

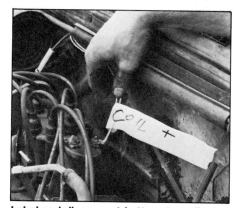

Label and disconnect battery-to-coil wire. This is the small wire at the positive (+) coil terminal. In the case of an HEI system, disconnect the three-wire connector.

fins. They fold up like an aluminum can under an elephant's foot and must be kept straight for efficient cooling. They could also double as a cheese shredder, and your knuckles could be the cheese! A piece of cardboard taped to the backside of the radiator core will protect it and your knuckles.

Remove the fan shroud after the radiator is out. If the radiator is in other than perfect shape, take it to a shop for reconditioning. It will then be ready for installation when the engine is. Otherwise, store the radiator where it won't be damaged; the trunk is a good choice, but be sure all the coolant has drained.

Fan—The fan can be left on the engine, but it isn't worth the risk of damaging it during removal. There are two types of fans: fixed and clutch drive, or *viscous* drive. The fixed fan bolts directly to the water-pump-shaft flange. Remove these bolts to free the fan. The clutch-drive is more difficult to remove. Four bolts hold the fan to the viscous-drive unit; their heads face towards the pump. Don't remove these four bolts. Remove the four bolts that thread into the water-pump flange—like a fixed fan. Remove these four bolts and the fan and clutch come off as a unit. The pulley and belts then come free also. Store the unit with *blades down and viscous-drive up* so fluid won't drain out and ruin the clutch.

Air Cleaner—Unbolt the air cleaner and remove it. Remove hoses, such as vacuum hoses, breather hose or hot- and fresh-air ducts so they stay with the air cleaner. The breather is usually at the rear of the air cleaner, the hot-air and fresh-air ducting is attached to the air-cleaner snorkle. At

reassembly they tend to fall back to their original positions, but don't forget to mark them. Wrap masking tape around the hoses, making a "flag" that can be written on. Make notations with a felt-tip or ball-point pen. Some sketches or a few snapshots taken from different angles can also prove helpful when installation time comes.

Throttle Linkage—With the air cleaner out of the way, the throttle linkage can be disconnected at the firewall arm—if you have the rod-type linkage. Leave the carburetor end alone. Pull out the cotter key with a pair of pliers, leaving some linkage parts attached to the carburetor, others to the fire wall. Disconnect a cable-type throttle linkage at the carburetor and the manifold mounting bracket. Lay or tie the cable back against the firewall so it'll be out of the way.

Depending on how your vehicle is equipped, there may be two other linkages to disconnect at the carburetor: With an automatic transmission, disconnect the kickdown linkage at the carburetor. Wire it to the firewall. Cruise control will have a cable similar to a throttle cable. Disconnect it and tie it out of the way.

Miscellaneous Wires—Disconnect and label the small wire from the ignition-coil positive post. Or, disconnect an HEI-distributor three-wire connecter. Up front, next to the thermostat housing on pre-'68 big-blocks, is the water-temperature sending unit. Later engines mount the water-temperature sender in the left cylinder head. Pull off the wire and label it.

A/C Compressor—One big item that has to come off is the air-conditioning

(A/C) compressor. The best way to remove it is as a unit; with hoses and bracket. The A/C system contains freon gas under high pressure.

CAUTION: Escaping freon gas is dangerous. When it escapes and expands, it goes through an extreme temperature drop. Escaping freon will cause frostbite. *Wear gloves and safety*

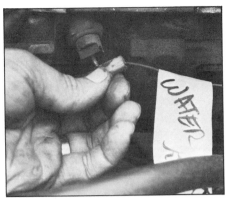

Early water-temperature senders are in intake manifold; post-'68 engines such as this one have sender in left head, between center exhaust ports. Mark wire before pulling it off.

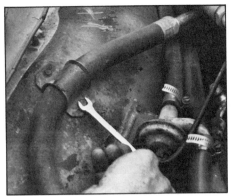

To get more slack in A/C lines, unbolt them from inner fenders.

After unbolting A/C compressor, tie it to an inner fender or the radiator support. Coathangers work well for this job. Another way to support compressor is at the end of a rope strung out of engine compartment and over fender. If you use this method, put some rags under the rope and lines to protect the body paint. Avoid disconnecting compressor lines; it's dangerous and will require recharging.

glasses when working on a charged A/C system. Otherwise you risk frostbite and blindness. Once reinstalled, the A/C system must be recharged or it will not work. Recharging an A/C system requires special equipment, including a vacuum pump.

Obviously, it is preferable not to discharge the A/C system if it can be avoided. However if it's not possible because one or both lines must be disconnected, do it right.

After putting on safety glasses and gloves, loosen the hose fitting marked DISCHARGE or D—it's on the low-pressure side of the compressor, or the line coming from the evaporator at the firewall. Wrap the fitting and the wrench with a rag. Loosen the fitting slowly until you hear hissing, then stop. The sound is escaping freon. Avoid getting in the path of the freon.

When the hissing stops, loosen the line another 1/4 turn. Wait for the hissing to stop, then disconnect the line. Keep the rag wrapped around the fitting just in case. To keep the A/C lines and compressor free of contamination, tape over the ends of hoses and compressor fittings. A sandwich bag tapped over the end of each hose works well.

Although there are different compressor styles and mounting arrangements, the basics are the same. Here are two typical removal procedures:

If the compressor is on the right side of the engine compartment—same side as the evaporator case—you're in luck. You'll be able to remove and position the compressor and its lines intact so they won't interfere with removing the engine.

After loosening and removing the drive belt, remove the A/C compressor/bracket from the engine. The compressor bracket mounts to the engine using two exhaust-manifold bolts at the side and two bolts that thread into the front of the left cylinder head. The top of the bracket is stabilized at the top by a long, slotted bar secured by the upper water-pump bolt. Loosen the adjusting nut and remove that nut and lock washer from the water-pump end. This special bolt looks like a stud with a hex head at the center. Once the compressor and brackets are loose, swing the compressor—lines and all—over the fender. Wire the compressor under the wheel-well opening so it's out of your way.

If the compressor is on the left side, you'll follow a different procedure. Before loosening the bracket and compressor, trace the routing of the two A/C lines to the evaporator. They are routed through clips attached to the inner fender panel. With these clips removed you should gain enough slack to move the compressor out of the way without disconnecting the lines.

Compressors mounted on the left side use a two-piece bracket, one piece goes between the front of the compressor and the cylinder head. The second piece attaches between the rear of the compressor and the intake manifold.

Start removal by loosening tension and pulling off drive belt. Remove the compressor from its brackets. Leave the rear bracket attached to the intake manifold and the front bracket bolted to the head. You can remove them after the engine has been lifted out.

Power-steering-pump *remote* reservoir bolted to top of alternator bracket on this 454 pickup engine. Reservoir and gound-strap should be removed from bracket and set aside. Keep lines connected and support reservoir upright. The same goes for the integral-type pump/reservoir.

Next, lower the compressor out of the way to the front of the engine compartment and support it with some wire. A coat hanger works great.

Power Steering—The power-steering pump mounts low on the left front of the engine. Both early- and late-model power-steering pumps are removed in a similar fashion. Two bolts attach the bracket to the front of the block and two retain it to the side. The front bolts are placed vertically. The top bolt is in plain sight, but the bottom one is accessed through a hole in the bracket with a socket wrench. On the side, the bolts are in a horizontal pattern.

Remove fuel-inlet hose from fuel pump. Pinch hose clamp open and slide it back. Twist hose to break it loose and pull it off pump.

Never rely on a jack to support a vehicle, only to lift it. Instead, use jackstands for support while you work underneath. Use 1500-lb-minimum-capacity stands securely placed on the ground and under the car frame. Block rear wheels so the vehicle cannot roll off of stands. Jostle vehicle to make sure it is secure before getting underneath.

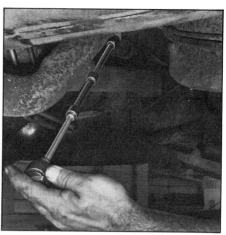

Penetrating oil on bolts or nuts and a six-point socket and ratchet with a long extension—or a bunch of short ones—are needed to disconnect exhaust pipes. A flex socket or universal at the socket may also be required.

The hose problem is similar to that of the A/C compressor, with two exceptions: Power-steering fluid—similar to ATF—is not dangerous; and special equipment is not required to refill—as opposed to recharge—a power-steering system. However, there's no need to disconnect the lines and create an oily mess, or risk contaminating the power-steering system. Instead, unbolt the pump bracket and place pump and bracket to the side. Wire it upright so fluid won't spill.

Air Pump—Later Air Injection Reactor (A.I.R.) pumps are mounted with the alternator on the right side. Remove this type pump later with the alternator. If you have the early big-block with an A.I.R. pump mounted by itself, remove it next. First disconnect the two rubber hoses at the two check valves. Leave the steel A.I.R. manifolds on the exhaust manifolds. There is no need to remove them. Disconnect the remaining hose at the carburetor.

Dismount the air pump itself by unbolting its bracket from the two upper water-pump bolts and from the head via the bolt below the unit. This way you can lift air pump, bracket, diverter valve, dump valve, rubber hoses and check valves off as a unit. Make triple-sure you label all the hoses and their connections clearly.

Idler Pulley—A few big-blocks have an accessory-drive idler pulley. All mount similarly, so they all are removed similarly. Three bolts hold the pulley assembly to the head. Remove these three bolts and remove the pulley-and-bracket assembly.

Fuel Pump—Although it is mounted low on the engine, the fuel pump and hoses are best reached from above. *Before disconnecting any fuel lines,* make sure there are no open flames, such as a gas water-heater pilot light, near your work area. Disconnect the fuel-tank-to-pump hose from the pump. Squeeze a spring-type hose clamp open with pliers and slide the clamp back. Otherwise loosen it with a screwdriver. Twist the hose to break it loose from the pump nipple and slide it off. Leave the pump on the block, but push a 3/8-in. bolt into the hose and secure it with a clamp. The bolt must have an unthreaded shank in the hose or gasoline will leak past the threads.

Alternator—Leave the alternator in place if your car has an automatic transmission. This assumes you want to use the starter motor to turn the crank when it comes time to unbolt the converter from the flexplate. The alternator is needed to complete the starter circuit. Otherwise, remove the alternator, page 22.

UNDERNEATH

With most of the work complete "upstairs," it's time to slide underneath. First, the car must be raised and safely supported. Start by blocking both rear tires with 4x4-inch blocks or something similar. Roll the floor jack under the car and position

the lifting pad under a frame side rail—at a flat spot so the jack will not slip—about 1/3 of the way toward the rear of the car. On Corvettes, place jack under a frame rail, about 5 in. in front of the door. Raise one side of the car at a time and immediately place a jackstand under the frame to take the load. Do not put stands under the front suspension.

When properly supported the car will be rock-steady. **Never rely on the floor jack, or any other lifting device, to support a vehicle.** Use jacks only for lifting, not supporting—they fail more often than you'd think. These failures are written about in the obituaries. When supporting a car with jack stands, check their load rating, then place them securely. **Do not use bricks, cinder or cement blocks.** They will crumble when subjected to high point loading. Whatever type of support you use, make sure the supports are spaced wide apart so the car won't fall sideways.

Exhaust—First job under the car is to disconnect the exhaust system at the manifolds. Each head pipe butts against an exhaust-manifold flange. The pipes are pulled tight to each manifold by nuts and a collar that slips over two or three studs. The nuts rust solid, making removal difficult. Your best friends here are "rust-busting" penetrating oil and a long, stout, 1/2-in.-drive ratchet and extension turning a six-point deep socket. If you don't have a straight shot at the nuts,

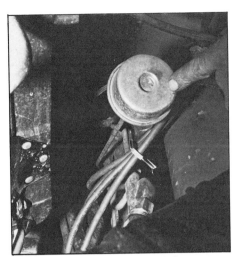

Early fuel-evaporation (EFE) system operated heat-risers have a vacuum can mounted above right-side exhaust manifold. Disconnect vacuum-can rod at heat riser.

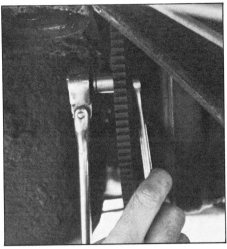

Flexplates are joined to their converters by three nut-and-bolt combinations; you'll have to turn the crank to gain access to all three. If weld-nuts are used, you only need one wrench to remove the bolts. Otherwise, you'll need two like this. If you used the starter to turn the crank, remove the battery and alternator after the bolts are out.

use a flex socket or U-joint. Give the penetrating oil time to soak in before you attempt to loosen the nuts.

Between the right exhaust manifold and header pipe is a heat-riser valve. Early big-block heat-riser valves are operated by a bimetal spring. This spring usually breaks, leaving the damper free to rattle. Replace the heat-riser-valve assembly if the spring is broken.

Later engines with *early fuel-evaporation* (EFE) systems have a vacuum-operated heat-riser valve. The vacuum can is mounted near the rocker cover on a bracket attached to the exhaust manifold. A rod extends from the vacuum can to the heat-riser-valve actuating arm. The rod is secured by a press-on connector. Pry the rod off with a screwdriver. The rest of the EFE system can be removed with the exhaust manifold once the engine is out.

The heat-riser valve—regardless of type—can now be removed with the exhaust pipe.

Once each nut breaks free, back it off a half turn or so. Give the nut another shot of penetrant and retighten it to loosen the rust and dirt that has "balled up" in the threads. Repeat this process each time the nut gets tight and you will avoid stripping threads. Once unbolted, slip the collar off the studs to separate the pipe from the manifold. Do both sides, and let the pipes hang down out of the way.

Corvettes require an extra step in disconnecting the exhaust. The mani-

folds and pipes will separate *only* after the front stud on each manifold has been removed—usually a huge headache. If more *slack* is needed to free the header pipes, loosen the exhaust system on its hangers all the way to the rear of the car. A combination of this and raising the engine slightly should do the job.

Some mechanics like to replace the studs and nuts with capscrews. Removing the studs is easy to do when the engine is out of the chassis. If you change to capscrews, make sure you use *anti-seize* compound on the threads before installing them.

Oil-Pressure Sender—Label and remove the wire from the oil-pressure sending unit. It's at the oil-filter adapter.

Converter to Flexplate—There's nothing to disconnect inside a manual-transmission bellhousing. On automatics and manual-transmission housings, you do need to remove the inspection cover. It closes off the bottom of the bellhousing. Remove the bolts and the cover. The torque converter and flexplate or the clutch and flywheel will now be exposed.

With an automatic transmission, remove three bolts that secure the flexplate to the torque converter. Some flexplate connections use bolts threaded into loose nuts; others use welded-on *nut plates*. One bolt will be in sight, but the other two will be inaccessible. Before removing any

bolts, mark the flexplate and converter so they can be reassembled in the same position. Theoretically the two should mate in any one of three positions, but this is frequently not the case. So avoid the problem by marking them. Put a dab of paint on the flexplate by one of the three bolt holes and on the companion torque-converter drive flange.

Remove the first bolt with a 9/16-in. box-end wrench on the bolt head and a ratchet on the nut—if a loose nut is used. To reach the remaining nuts and bolts, the crankshaft must be turned.

To turn the engine, reattach the battery cables so the starter can do the heavy work. If you have a friend operating the starter while you are underneath, make doubly sure you understand each other's signals. If he engages the starter while you have a hand in the bellhousing, it could be a real disaster. Best instruct your helper to remove the ignition key after each nudge. Disconnect and remove the battery after you have the converter unbolted from the flexplate.

There is another way to turn the crankshaft. Use a 3/4-in. socket and a long breaker bar on the damper bolt. You can now turn the engine and torque converter. Actually, this method goes quickly, especially with two people—one to turn the crank and the other to remove the nuts and bolts. It's even quicker with the spark-plugs out.

If you are working on a frozen engine that won't turn, you can remove the flexplate connections with a *long* open-end wrench and—if loose nuts are used—a long, thin screwdriver. Both tools should be longer than 9 in. Use the wrench on the bolt while prying against the nut with the screwdriver so it can't turn. That's the way I had to disconnect one big-block.

Starter Motor—On Corvettes, the first step is to remove the heat shield. It covers the starter solenoid and the upper half of the starter motor. The heat shield is secured at the front by the support bolt that passes first through the shield, then the front support and finally into the block. At the rear of the starter, the shield is held in place by a nut that can be seen as you look at the starter from the rear. Remove these fasteners and the shield.

Remove bolts securing the starter. The first is in a bracket at the starter nose. Leave bracket on starter. The other three bolts—two in rare cases—thread into block at the rear. Watch out—the starter is heavy. Wire up starter or position it so it doesn't hang on its leads if you left them connected.

Off-the-car shot shows starter-support bracket. Leave it in place on starter.

Now for the electrical connections. You can disconnect them or leave them in place. If you want to remove the starter, so it can be cleaned and inspected, disconnect the wiring. Label the wires as usual, otherwise you'll never remember where they go. Once labeled, disconnect the wires from solenoid and return the nuts to their studs. Unbolt the front support at the block, then hold the starter in place with one hand. Move to the other end of the starter and remove the three bolts—two in rare cases—that extend vertically through the starter nose and into the block, or bellhousing with some truck engines.

Keep your head out from under the starter or you could be hurt. The starter may drop once the last bolt is out, so be careful—it's heavy.

If you elect to leave the starter leads connected, wire the starter to the frame or firewall, or alongside the exhaust pipe. Don't let it hang by its electrical connections! If you are interested in a fast rebuild and aren't going to clean the engine compartment or accessories completely—this method is for you.

Engine Mounts—On each side of the block is an engine mount. Although these rubber/metal pads support the engine, they can be unbolted while the engine rests on them. Each engine mount consists of two parts: one bolts to the engine block, the other bolts or is welded to the chassis. The two engine-mount parts are joined with a long through bolt. Remove the through bolt from each mount and

the mounts will separate as the engine is lifted. The parts that bolt to the engine can be removed after the engine has been lifted out.

Hold the through bolt with a box-end wrench and loosen the nut with a ratchet and socket. Let the wrench swing around against the block so you won't have to hold it. After the nut is off, slide the bolt out. Remove the other bolt. Some mounts use a weld-nut. This being the case, all you have to do is loosen and remove the through bolt—the nut will stay put.

Clutch Linkage—There's one more job to do underneath, if you're working with a manual transmission: The clutch-linkage cross shaft must be removed. Start by removing the clutch-fork return spring. Next, remove the clutch-fork-to-cross-shaft pushrod by disconnecting it at the cross shaft. The spring clip that retains the pushrod to the cross shaft is best removed with needle-nose pliers.

Turn your attention to the top of the shaft. Disconnect the pushrod that is retained similarly. You may have to do this from above. Now you can remove the cross shaft.

If the bracket at the chassis end of the cross shaft is slotted at the pivot, loosen the nut that retains the pivot. Slip the end of the pivot out of the bracket. Pull the cross shaft off its pivot at the engine and remove it. If the bracket is not slotted, unbolt the bracket, then remove the cross shaft.

There is no more work to be done from underneath, so lower the car.

Using a ratchet and socket with a long extension to remove engine-mount through bolt. A helper is below with a box-end wrench on the nut. If you are doing this job yourself, let wrench turn against block and hold itself.

Removing an engine with the car raised only cheats you out of cherry-picker-arm travel.

Alternator—Like the starter, the electrical connections can be left intact. Simply remove the alternator, set it aside and secure it, so it doesn't hang on its wiring harness. If you want to remove the alternator so it can be cleaned up and inspected, label and disconnect the electrical connections. Loop the bundle of wires out of your way.

On all but a Corvette, loosen the through bolt at the alternator pivot and unbolt the adjusting bracket at the block. The support bracket curves around the corner of the cylinder head, under the alternator. Two bolts at front and two at the side must be removed. Both brackets and alternator can then be removed.

When alternator and air-pump are mounted together, there's no guarantee alternator will be on top. Bracketry on this 366 truck engine is a little different from the standard passenger car/pickup. Regardless, alternator and air pump are removed as a a unit.

Early alternator is mounted by itself. Label and disconnect the two alternator leads, then remove alternator from engine compartment. Or, leave electrical connections intact and set alternator aside *in the engine compartment.* Don't let alternator hang from its electrical leads.

Later alternator mounted singly pivots on a long stud that threads into right cylinder head. Remove nut from stud and wide adjusting bracket, then slide alternator off.

Battery tray works well for storing alternator in this engine compartment. Alternator stud is still sticking out from the head. Remove it so you don't get stabbed.

Remove battery-ground cable from engine.

How to do it if your engine doesn't have lifting plates: Run a bolt into an accessory-mounting hole in the left cylinder head and the rear of the other. Use a large-diameter flat washer under each bolt head to prevent the chain link from coming off.

Although the Corvette-alternator bracket looks considerably different, it's no more difficult to remove. The big difference is at the front of the head. Remove the bracket bolt closest to the center of the engine. At the side there are still two bolts to remove.

Ground Straps—Look for the engine ground straps. You should find two at the rear—flat woven-steel cables—between the corners of each rocker cover and the firewall. Their ends are secured under rocker-cover bolts. Unbolt the ground straps and let them hang from the firewall. Thread the bolts back into their rocker covers—finger tight.

Remove the battery-ground cable while you are at it. You'll find it bolted to the side of the right cylinder head, at the front. Keep the bolt with the cable.

Hooking Up—Now you're ready to hook up the engine hoist. Position the hoist so its hook centers over the carburetor. Attach the lifting chain or cable diagonally across the engine. If you're lucky you'll find an engine lift plate at both the front and rear of the intake manifold. If you've got 'em, use 'em. If not, use power-steering bolts or the like to bolt the chain to the front of one head and the rear of the other.

To keep the chain from coming off the bolts, place large-diameter washers between each bolt head and the chain. Tighten the bolts.

Properly attached, the chain should pull tight about 5-in. above the carburetor, at its center. Hook onto the chain at its center or a little forward so the engine will come up level or slightly up at the front.

Before you lift, position the floor jack securely under the transmission. The transmission must be supported before the engine can be pulled from it. Otherwise clutch or converter damage may result.

Bellhousing Bolts—All bellhousing bolts can be reached from above—but you'll have to stretch for a couple of them. Remove *all six or seven.* If you drop a bolt, don't worry. You can retrieve it after the engine is out.

A common mistake is overlooking the bolts at the bottom corners of the bellhousing. This is especially true on the right side, where the bottom bolt is hidden by the exhaust. Just make sure they are all out.

Lift Engine—Double-check that everything that must be disconnected or removed is. Although you may have been working alone up to this point, now's the time to summon a couple of friends. Engine removal will be safer and go quicker if you have help. One person operates the hoist,

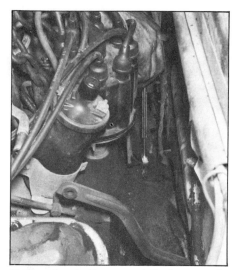

Bellhousing bolts are relatively easy to remove—except from a Corvette. Most bellhousing bolts can be reached from above, however you may have to go underneath to get those at the bottom corners of the bellhousing. Double-check to make sure you removed them all.

Ideal setup: one watching underneath, one guiding engine from above, and the third operating hoist. As engine is raised, watch for hang-ups. If you feel resistance, stop and investigate.

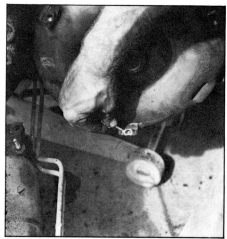

If you haven't already done so, mark flexplate and torque converter immediately after pulling engine.

ready to raise or lower on command. The other two work at the sides of the engine and guide it out. As the engine is raised, the one with the floor jack on his side must keep the jack centered under and in contact with the transmission. If you are alone, you'll have to do all three jobs yourself, doing one incremental step at a time.

An engine will disengage from an automatic transmission with little trouble. On the other hand, a manual transmission has a long input shaft that splines into the clutch disc and pilots into the center of the crankshaft. This means the engine must be moved forward several inches before it will disengage from the input shaft.

Regardless of the type of transmission, if more than a couple of tugs are required to pull the engine free of the transmission, something is still attached. Find the problem before proceeding.

Once the transmission is cleared, raise the engine. Keep an eye out for any odd wire still connected. If the engine hangs up on anything, **stop.** Find the problem before you continue to lift. If you don't, you'll only risk damage to whatever is still attached to the engine.

Mark Converter & Flexplate—If you haven't *indexed* the converter and flexplate, do it now. As soon as the engine is out, and *before* you rotate the crank or converter, mark both

converter and flexplate at the *six-o'clock* points with light-colored paint as shown. Don't omit this step or you may have assembly or balance troubles later.

Loose Ends—Before the neighborhood kids abscond with the hardware off your engine to repair their skateboards, put it away. Get every piece of hardware, tool and part into the garage, basement or trunk. Many parts—like the alternator—can go in the trunk. Use labeled containers for each accessory and its nuts, bolts and washers. Separate them. Four bolts in a marked plastic bag are four bolts you don't have to hunt for later. Twenty miscellaneous bolts in a jar create a time-wasting problem.

Unbend a few more coathangers to hang the exhaust pipes and transmission. **Don't** remove the floor jack until the transmission is supported. Use a pipe or 2x4 to bridge the engine compartment. Run the wires through the bellhousing bolt holes and over the pipe or 2x4.

Skip the next chapter for now and go straight to engine-teardown chapter, page 49. You can start stripping the engine while it hangs from the engine hoist. Have boxes, cans or bags for storing parts as they come off the engine.

When you're done for the day read Chapter 3 on Parts Identification & Interchange. It can save you some money.

3 Parts Identification & Interchange

Besides the satisfaction of rebuilding your own engine, saving money is another big reason for doing it. And replacing badly worn parts with good used parts is one of the best ways you can save—not having to buy new pieces at new-part prices.

The key to successful rebuilding with used parts is knowing what parts interchange—what parts from one engine will go into another. For example, if you are rebuilding a 396, it may be helpful to know that the crankshaft from *any* 366, 396, 402 or 427 will work. With a list of all the interchange possibilities at hand, you can save a lot of time and money if the need arises to visit your nearby "preowned auto-parts store"—junkyard.

Don't worry about your parts-hunting expedition turning into a wild-goose chase. If you've done any parts swapping with other engines, you'll be amazed at how many big-block-Chevy parts interchange. Besides being very strong and durable, these engines share an incredible number of parts. You should find parts swapping easy.

Two skills must be developed before you can swap parts competently: First, you must be able to make positive identification of the parts being interchanged, both in your engine and those at the junkyard. Second, you must know which parts will interchange directly or those that will require minor work. And you must keep in mind how you intend to rebuild your engine—for standard performance or for high performance. Building a high-performance engine—such as one for racing—requires specific parts for the specific application. If you are after more than standard *factory* performance from your big-block Chevy, you'll also need HPBooks' *How to Hotrod Big-block Chevys.*

Vehicle Identification Number— Before you can swap parts, or even order replacements, you must identify your engine. The first ID *clue* is the *vehicle identification number,* or VIN.

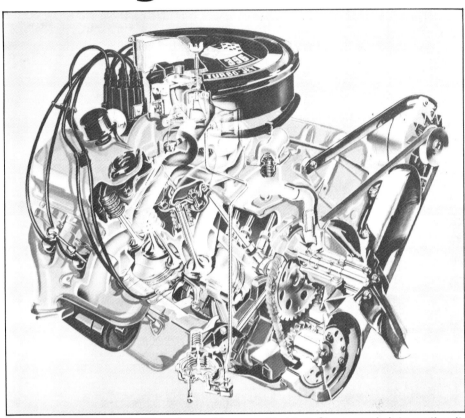

Cutaway of 325-HP 396. Although a part from one big-block Chevy may bolt to another, it doesn't mean it will work. Read this chapter carefully if you plan on interchanging any parts on your big Chevy. Photo courtesy of Chevrolet.

Chassis built prior to 1968 have the VIN stamped on a plate attached to the left-front door pillar, or on the 'Vette, it's on the instrument-panel brace at the right. On '68 and later models you'll find the *VIN plate* at the extreme left at the base of the windshield atop the instrument panel.

Like most part and ID numbers, the VIN looks hopelessly complicated at first, but decodes easily. The first digit of the *alphanumeric* "number" tells which GM Division—Chevrolet, Olds, Pontiac, . . . —built the car or truck.

In the second digit is a letter denoting the car line as Impala, Nova, Camaro or whatever. Body Style—station wagon, coupe or four-door sedan—is identified by the third and fourth digits. This is interesting, but you really want to know the

Engine Type, as shown by the next letter—the fifth digit.

When the fifth digit is decoded, using the table on page 28 or the one at your Chevy parts department, you will get a *regular production option* (RPO) number, plus engine displacement and number of carburetor barrels. The number will look something like this: LS-4 454 cu. in. 4-bbl. Look up the LS-4 designation to get a detailed explanation of what's inside the powerplant.

The remainder of the VIN gives the Model Year, a letter code for the Assembly Plant and, finally, the *sequence number*—position of the car on the assembly line.

Using RPO Numbers—Before running to a dealer to buy goodies to fit, say an L-35, you must realize that RPO numbers don't mean much to a

Vehicle Identification Number

1	L	57	H	3	S	100001

Division (1-Chevrolet)
Car Line & Series
Basic Body Style
Engine Type
Model Year (1973)
Assembly Plant
Sequential Number

Car Line and Series

C—Deluxe (Chevelle)/El Camino
D—Malibu/El Camino Custom
E—Laguna
G—Malibu Estate
H—Monte/Carlo/Laguna Estate
K—Bel Air
L—Impala
N—Caprice Classic/Caprice Estate
Q—Camaro
S—Camaro Type LT
V—Vega
X—Nova
Y—Nova Custom
Z—Corvette

Basic Body Style

39—Sport Sedan (Chevrolet)
47—Coupe (Chevrolet)
57—Sport Coupe (Chevrolet/Monte Carlo)
57/Z03—Landau Coupe (Monte Carlo)
57/Z76—S Coupe (Monte Carlo)
67—Convertible (Chevrolet/Corvette)
69—4-Door Sedan (Chevrolet/Nova)
29—4-Door Sedan (Chevelle)
37—Sport Coupe (Chevelle/Corvette)

35—2-Seat Wagon (Chevrolet/Chevelle)
35/AQ4—3-seat Wagon (Chevelle)
45—3-seat Wagon (Chevrolet)
87—Sport Coupe (Camaro)
17—Hatchback Coupe (Nova)
27—Coupe (Nova)
80—El Camino

Engine Type

Ordering Code	Disp. (cu.in.)	Carb. (bbl.)
R-LF6	400	2
Y-LS4	454	4
X-LS4	454	4

Assembly Plant

A—GMAD—Lakewood
B—GMAD—Baltimore
C—GMAD—Southgate
D—GMAD—Doraville
F—Chevrolet—Flint
J—GMAD—Janesville
K—GMAD—Leeds
L—GMAD—Van Nuys
N—Chevrolet—Norwood
V—G.M. TRUCK—Pontiac

R—GMAD—Arlington
S—Chevrolet—St. Louis
T—GMAD—Tarrytown
U—Chevrolet—Lordstown
W—Chevrolet—Willow Run
Y—GMAD—Wilmington
Z—GMAD—Fremont
1—GM of Canada—Oshawa
2—GM of Canada—St. Therese

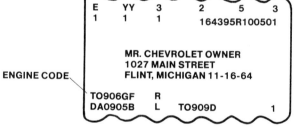

1965 PROTECT-O-PLATE

E	YY	3	2	5	3
1	1	1	164395R100501		

MR. CHEVROLET OWNER
1027 MAIN STREET
FLINT, MICHIGAN 11-16-64

ENGINE CODE

| TO906GF | R | | |
| DA0905B | L | TO909D | 1 |

1966-72 PROTECT-O-PLATE

D	AA	166457F123456	R
F1210HC		EU0212G	P
C7E03		3133311	

ENGINE CODE

MR. CHEVROLET OWNER
1027 MAIN STREET
FLINT, MICHIGAN 11-16-67
USA

Who is "Mark," anyway?

Chevrolet engineers and enthusiasts refer to the big-block as "Mark" engines. There have been several "Marks."

Mark 1—348/409 CID series
Mark 2—427 CID "Mystery" engine first used at Daytona in 1963
Mark 3—never released
Mark 4—366/396/402/427/454—the familiar Turbo-Jet engines

ENGINE RPOs IN CHEVROLET PASSENGER CARS

Year	Engine	RPO	Car Line	Year	Engine	RPO	Car Line
1965	396	L35	Chevrolet	1979	427	LS1	Chevrolet
		L37	Chevelle			L36	Chevrolet, Corvette
		L78	Chevrolet, Corvette			L68	Corvette
1966	396	L35	Chevrolet, Chevelle SS, El Camino			L72	Chevrolet
						L71	Corvette
		L34	Chevelle SS, El Camino			L88	Corvette
		L78	Chevelle SS, El Camino	1970	402	L34	Chevelle, Camaro, Nova
	427	L36	Chevrolet, Corvette			L78	Chevelle, Camaro, Nova
		L72	Chevrolet, Corvette			LS3	Chevelle, Monte Carlo
1967	396	L35	Chevrolet, Chevelle SS, El Camino SS, Camaro SS		454	LS4	Chevrolet
						LS5	Chevelle, Monte Carlo, Chevrolet, Corvette
		L34	Chevelle SS, El Camino, Camaro SS			LS6	Chevelle, Camaro
		L78	Chevelle SS, El Camino, Camaro SS			LS7	Chevelle, Corvette
	427	L36	Chevrolet, Corvette	1971	402	LS3	Chevrolet, Chevelle, Monte Carlo, Camaro
		L68	Corvette		454	LS5	Chevrolet, Chevelle, Monte Carlo, Corvette
		L72	Chevrolet			LS6	Chevelle, Monte Carlo, Corvette
		L71	Corvette	1972	402	LS3	Chevelle, Monte Carlo, Camaro, Chevrolet
1968	396	L35	Chevrolet, Chevelle SS, Camaro SS, El Camino SS		454	LS5	Chevrolet, Chevelle, Monte Carlo, Corvette
		L34	Chevelle SS, El Camino SS, Camaro SS	1973	454	LS4	Chevrolet, Monte Carlo, Chevelle, Corvette
		L78	Chevelle SS, El Camino SS, Camaro SS	1974	454	LS4	Chevrolet, Monte Carlo, Chevelle, Corvette
	427	L36	Chevrolet, Corvette	1975	454	LS4	Chevrolet, Monte Carlo, Chevelle (except Laguna)
		L68	Corvette	1976	454	LS4	Chevrolet
		L72	Chevrolet				
		L71	Corvette				
		L88	Corvette				
1969	396	L66	Chevrolet				
		L35	Chevelle, Camaro				
		L34	Chevelle, Nova, Camaro				
		L78	Chevelle, Nova, Camaro				

Reference this chart to determine what RPO big-block was installed in what chassis.

parts man. RPO numbers are assigned by the engineering department. Although a parts man may be able to decipher RPO numbers, or may understand what you are talking about, he really needs the model year and body style. This even applies to numbers that are part of the Chevy jargon—L-88, LS-6, ZL-1 . . .

For example, L-88 decodes to nothing more than **special engine**, so stick with the model year and body style, and refer only to the RPO for your own information.

Protect-O-Plate—If you are dealing with a 1965—'72 model-year vehicle, you may have an additional information source—the *Protect-O-Plate*. I say "may" because many owners discard or lose their Protect-O-Plates after the warranty period, or they are simply lost. The first place to look is in the glove compartment.

Embossed like a credit card, the metal Protect-O-Plate is used for warranty purposes. Its codes contain the specifics of how each car was equipped at the factory. You'll also find the original owner's name embossed by the dealer on a special plastic strip.

Engine-wise, what you can learn from the Protect-O-Plate are codes for engine-RPO number, day the engine was built and in what plant. The carburetor-manufacturer code is also given. Additionally, the Protect-O-Plate can provide: rear-axle ratio, options such as air conditioning and power brakes, transmission type and where manufactured. Because all the pages in this chapter would be needed to cross-reference the information reflected on all Protect-O-Plates, I'll just include engine information.

Two numbers are significant: the VIN at the top center, and the *Engine Production Code* number at the left

edge. The VIN should agree with the one on the instrument panel or door pillar. The Engine Production Code number provides some additional engine information not found in the VIN. It starts with a letter code for the plant where the engine was built.

The next four digits in the Engine Production Code number supply the month and day of manufacture. The engine suffix follows. Two or three letters make up the engine suffix. First described is the transmission originally bolted to that particular engine. If the application is for a taxi or police car, it is noted, as are engine-performance options. From 1970 on, the engine's rated horsepower was provided.

By matching numbers with the information in this book, or by using the VIN, you can identify your engine. Let's use as an example an Engine Production Code number

PASSENGER-CAR SERIES NUMBER & SUFFIX

1965 396 CHEVROLET
Man. Trans.	IA,LF
Man. Trans. and Spec. High Perf.	IE
Man. Trans. and Transistor Ignition	IC
Powerglide	IG,LB
Powerglide and Transistor Ignition	II
Turbo. Hydramatic	IV,LC
Turbo. Hydramatic and Transistor Ignition	IW

1966 396 CHEVROLET
Man. Trans.	IA
A.I.R.	IB
Powerglide	IG
3-Speed Auto. Trans.	IV
A.I.R. with Powerglide	IC
A.I.R. and 3-Speed Auto. Trans.	IN

427
Man. Trans.	IH
Special High Perf.	ID
3-Speed Auto. Trans.	IJ
A.I.R.	II
High Perf. with A.I.R. and 3-Speed Auto. Trans.	IO

396 CHEVELLE
Man. Trans.	ED
High Perf.	EF
Special High Perf.	EG
A.I.R.	EH
High Perf. and A.I.R.	EJ
Powerglide	EK
High Perf. and Powerglide	EL
Powerglide and A.I.R.	EM
High Perf., Powerglide and A.I.R.	EN

1967 396 CHEVROLET
Man. Trans.	IA
A.I.R.	IB
Powerglide	IG
3-Speed Auto. Trans.	IV
A.I.R. and Powerglide	IC
A.I.R. and 3-Speed Auto. Trans.	IN

427
Man. Trans.	IE,IH
Special High Perf.	ID
3-Speed Auto. Trans.	IJ,IS
A.I.R.	II,IX
Special High Perf. and A.I.R.	IK
A.I.R. and 3-Speed Auto. Trans.	IF,IO

396 CHEVELLE
Man. and Powerglide Trans.	ED
High Perf.	EF
Special High Perf.	EG
A.I.R.	EH
High Perf. and A.I.R.	EJ
Powerglide	EK
High Perf. and Powerglide	EL
Powerglide and A.I.R.	EM
High Perf., Powerglide and A.I.R.	EN
3-Speed Auto. Trans.	ET
High Perf. & 3-Speed Auto. Trans.	EU
3-Speed Auto. Trans. and A.I.R.	EV
High Perf., 3-Speed Auto Trans. & A.I.R.	EW
Special High Perf. and A.I.R.	EX

CAMARO
High Perf.	EI
High Perf. and 3-Speed Auto. Trans.	EQ
High Perf. and A.I.R.	EY
Special High Perf.	MQ
A.I.R. Special High Perf.	MR
Man. and A.T. Trans.	MW
A.I.R.	MX
3-Speed Auto. Trans.	MY
3-Speed Auto. Trans. and A.I.R.	MZ

1968 396 CHEVROLET
Man. Trans.	IA
Powerglide	IG
3-Speed Auto. Trans.	IV
(Police) Man. Trans.	IK
(Police) Powerglide	IN
(Police) H.D. Clutch	IF

427
Police 3-Speed Auto. T.H. 400	IB
Man. Trans.	IE-IH
Special High Perf.	ID
3-Speed Auto. Trans.	IJ-IS
(Police) Man. Trans.	IC

396 CHEVELLE
Man. Trans.	ED
High Perf.	EF
Special High Perf.	EG
Powerglide	EK
High Perf. and Powerglide	EL
3-Speed Auto. Trans.	ET
High Perf. and 3-Speed Auto. Trans.	EU

396 CHEVY II
Man. Trans.	MX
Man. Trans.	MQ
3-Speed Auto., T.H. 400	MR

396 CAMARO
Special High Perf.	MQ
High Perf. and 3-Speed Auto. Trans.	MR
Special High Perf. and Aluminum Heads	MT
Man. and AT Trans.	MW
High Perf.	MX
3-Speed Auto. Trans.	MY

1969 396 CHEVROLET
2-Bbl Carb. Man. Trans.	JN
(Police) 2-Bbl. Carb. Auto. Trans. T.H. 400	JO
(Police) 2-Bbl. Carb.	JP
2-Bbl Carb. Auto. Trans. T.H. 400	JQ
Man. Trans.	JT
(Police) Man. Trans.	JR

427
High Perf.	LA
4-Bbl. Carb.	LB
High Perf. Auto. Trans. T.H. 400	LC
Special High Perf. Man. Trans.	LD
4-Bbl. Carb. Auto. Trans. T.H. 400	LE
(Police) High Perf. Auto. Trans.	LF
(Police) High Perf. Man. Trans.	LG
High Perf. Man. Trans.	LH
3-Speed Auto. Trans. T.H. 400	LI
(Police) 4-Bbl. Carb. Auto. Trans. T.H. 400	LJ
(Police) 4-Bbl. Carb. Man. Trans.	LK
Special High Perf. Auto. Trans. T.H. 400	LS
(Police) Man. Trans.	LY
(Police) High Perf. Man. Trans.	LZ
Man. Trans.	MA
(Police) High Perf. Man. Trans.	MB
High Perf. Man. Trans.	MC
Special High Perf. Man. Trans.	MD

396 CHEVELLE
Man. Trans.	JA
High Perf.	JC
Special High Perf.	JD
High Perf. Auto. Trans. T.H. 400	JE
3-Speed Auto. Trans. T.H. 400	JK
Spec. High Perf. Auto. T.H. 400	KF
Man. Trans.	KG
3-Speed Auto. T.H. 400	KH
High Perf. Man. Trans.	KB
Man. Trans.	JV
Spec. High Perf. Man. Trans.	KD
Man. Trans.	KI

396 CHEVY II
High Perf.	JF
Special High Perf.	JH
High Perf. Auto. Trans. T.H. 400	JI
Special High Perf. 3-Speed Auto. Trans. T.H. 400	JL
3-Speed Auto. T.H. 400	JM
Powerglide	JU
Spec. High Perf. Man. Trans.	KA
Spec. High Perf.	KC
Man. Trans.	KE

396 CAMARO—396
Powerglide	JB
High Perf.	JF
3-Speed Auto. Trans. T.H. 400	JG
High Perf. Man. Trans.	JH
High Perf. Auto. Trans. T.H. 400	JI
Aluminum Heads Man. Trans.	JJ
Special High Perf. Auto. Trans. T.H. 400	JL
Aluminum Heads Auto. Trans. T.H. 400	JM
Man. Trans.	JU
Spec. High Perf. Man. Trans.	KA
Spec. High Perf. Man. Trans.	KC
Aluminum Heads Man. Trans.	KE

1970 454 CHEVROLET
Man. Trans.	345 HP	CGV
Man. Trans. and Police	345 HP	CGS
Man. Trans. and Police	390 HP	CGT
Man. Trans.	390 HP	CGU

396 CHEVELLE
T.H. 400	350 HP	CTW
Man. Trans.	350 HP	CTX
T.H. 400	375 HP	CTY
Man. Trans., H.D. Clutch	350 HP	CTZ
Man. Trans.	375 HP	CKO
T.H. 400, Aluminum Heads	375 HP	CKP
Man. Trans., H.D. Clutch	375 HP	CKQ
Man. Trans.	375 HP	CKT
Man. Trans., H.D. Clutch	375 HP	CKU

454
Man. Trans.	390 HP	CRN
T.H. 400	390 HP	CRQ
T.H. 400	450 HP	CRR
T.H. 400 and Aluminum Heads	450 HP	CRS
Man. Trans.	390 HP	CRT
Man. Trans.	450 HP	CRV

396 NOVA
T.H. 400	350 HP	CTW
Man. Trans.	350 HP	CTX
T.H. 400	375 HP	CTY
Man. Trans., H.D. Clutch	350 HP	CTZ
Man. Trans.	375 HP	CKO
T.H. 400 and Aluminum Heads	375 HP	CKP
Man. Trans., H.D. Clutch	375 HP	CKQ
Man. Trans., Aluminum Heads	375 HP	CKT
Man. Trans., Aluminum Heads and H.D. Clutch	375 HP	CKU

396 CAMARO
T.H. 400	375 HP	CJL
Man. Trans.	350 HP	CJF
Man. Trans.	375 HP	CJH
T.H. 400	350 HP	CJI

1971 400 CHEVROLET
T.H. 350	300 HP	CLP
Man. Trans., Police	300 HP	CLR

454
Man. Trans., Police	365 HP	CPG
Man. Trans.	365 HP	CPD

400 CHEVELLE
T.H. 350	300 HP	CLP
T.H. 400	300 HP	CLB
Man. Trans.	330 HP	CLA
4-Speed	300 HP	CLI
Man. Trans. Police	300 HP	CLR
Man. Trans.	300 HP	CLS

454
Man. Trans.	365 HP	CPA
Man. Trans.	365 HP	CPG
Man. Trans.	365 HP	CPD
Man. Trans.	460 HP	CPP
T.H. 400	460 HP	CPR

454 NOVA
T.H. 400	450 HP	CPT
Man. Trans.	450 HP	CPS

400 CAMARO
T.H. 400	350 HP	CLD
Man. Trans.	350 HP	CLC

1972 400 CHEVROLET
Man. Trans. (L-S3)	CLB
Man. Trans., Police (L-S3)	CLR
Man. Trans., w/A.I.R. (L-S3)	CTB
Man. Trans., Police w/A.I.R. (L-S3)	CTJ

454
T.H.	CPD
T.H. Police	CPG
T.H. w/A.I.R.	CRW
T.H. Police, w/A.I.R.	CRY

400 CHEVELLE
Man. Trans.	CLA,CLS
*T.H.	CLB
Man. Trans., w/A.I.R., Police and Taxi	CTA
T.H., w/A.I.R.	CTB
*T.H. w/A.I.R., Police and Taxi	CTJ
H.D. 3-Speed, w/A.I.R.	CTH

*Includes Monte Carlo

454
Man. Trans.	CPA
*T.H.	CPD
Man. Trans., w/A.I.R.	CRX
T.H., w/A.I.R.	CRW

*Includes Monte Carlo

400 CAMARO
Man. Trans.	CLA
T.H.	CLB
Man. Trans., w/A.I.R.	CTA
T.H., w/A.I.R.	CTB

1973 454 CHEVROLET
T.H., w/NB2	CWD
T.H., w/NB2	CWJ
T.H.	CWL
T.H.	CWK

454 CHEVELLE
Man. Trans. (4-Speed) w/NB2	CWC
T.H., w/NB2	CWD
Man. Trans. (4-Speed)	CWA
T.H.	CWB

1974 454 CHEVROLET
T.H. (M-40) w/4BC, Police	CWU
T.H. (M-40) w/4BC, NB-2, Police	CWW
T.H. (M-40) w/4BX, NB-2	CWY
T.H. (M-40) w/4BC, NB-2	CXA
T.H. (M-40) w/4BC, NB-2	CXB
T.H. (M-40) w/4BC, NB-2	CXC
T.H. (M-40) w/4BC	CXT
T.H. (M-40) w/4BC	CXU

454 CHEVELLE
4-Speed w/4BC	CWA
T.H. (M-40) w/4BC	CWX
T.H. (M-40) w/4BC, NB-2	CWD
4-Speed w/4BC	CXM
T.H. (M-40) w/4BC	CXR
T.H. (M-40) w/4BC	CXS

1975 454 CHEVROLET
	CXK,CXX
	CXL,CXY

454 CHEVELLE
	CXW

TRUCK SERIES NUMBER & SUFFIX

Engine type and application is identified by a code. Code is stamped on pad immediately forward of right cylinder head.

Model	Description	Code
1973—366		
M-65	A.I.R. w/M.T. 650-G	AAA
C-60-65-M-65-S-60	A.I.R. w/Manual Trans.	ADS
C-60-65-S-60	A.I.R. w/A.T. 540	ADU
C-60-S-60	A.I.R. w/A.T. 475	ADV
T-60-65	A.I.R. w/Manual Trans.	ADW
T-65	A.I.R. w/A.T. 540	ADX
C-60-65-M-65	L.P. Gas	AFT
T-60-65	L.P. Gas	AFU
C-60-65-M-65-S-60	Manual Trans. (exc. A.I.R., L.P. Gas)	AHK
C-60-65-S-60	A.T. 540 (exc. A.I.R.)	AHM
T-60-65	Manual Trans. (exc. A.I.R., L.P. Gas)	AHN
T-65	A.T. 540 (exc. A.I.R.)	AHR
M-65	M.T. 650-G (exc. A.I.R.)	AHS
C-60-S-60	A.T. 475 (exc. A.I.R.)	AJB
427		
C-65-M-65	A.I.R. w/M.T. 640-G, 650-G	AAB
C-65-M-65	A.I.R. w/Manual Trans.	ADZ
T-65	A.I.R.	AEA
C-65-M-65	Manual Trans. (exc. A.I.R.)	AHT
T-65	(exc. A.I.R.)	AHU
C-65-M-65	M.T. 640-G, 650-G (exc. A.I.R.)	AHW
1974—366		
C-60-65-M-65	L.P. Gas	AFT
T-60-65	L.P. Gas	AFU
T-60-65	Manual Trans. (exc. L.P. Gas)	AHA
C-60-S-60	A.T. 475	AHB
C-60-65-M-65-S-60	Manual Trans. (exc. L.P. Gas)	AHK
C-60-65-S-60	A.T. 540	AHM
T-65	A.T. 540	AHR
M-65	M.T. 650-G	AHS
427		
C-65-M-65	Manual Trans.	AKT
T-65		AKU
C-65-M-65	MT-640-G, 650-G	AKW
1975—366		
C-60-65, S60, M65	Manual Trans. (exc. Exh. Emission Control)	AHD
C65, M65	M.T. 640G, 650G, (exc. Exh. Emission Control)	AHH
C60-65, S60	A.T.540 (exc. Exh. Emission Control)	AHJ
C60-65, S60	A.T. 475 (exc. Exh. Emission Control)	AHL
T60-65	Manual Trans. (exc. Exh. Emission Control)	AHU
T60-65	A.T. 540 (exc. Exh. Emission Control)	AHW
C-60-65, S60, M65	Manual Trans. w/Exh. Emission Control	AHX
C65, M65	M.T. 640G, 650G, w/Exh. Emission Control	AHY
C60-65, S60	A.T. 540 w/Exh. Emission Control	AHZ
C60-65, S60	A.T. 475 w/Exh. Emission Control	ATA
T60-65	Manual Trans. w/Exh. Emission Control	ATB
T60-65	A.T. 540 w/Exh. Emission Control	ATC
427		
C65, M65	Manual Trans. (exc. Exh. Emission Control)	AKA
C65, M65	M.T. 640G, 650G (exc. Exh. Emission Control)	AKB
T65	Manual Trans. (exc. Exh. Emission Control)	AKC
C65, M65	Manual Trans. w/Exh. Emission Control	AKD
T65	Manual Trans. w/Exh. Emission Control	AKJ
C65, M65	M.T. 640G, 650G, w/Exh. Emission Control	AKH
1976—366		
C60-65, M65	Manual Trans. w/Air Brake (exc. Exh. Emission Control)	AHD
S60	Manual Trans. (exc. Exh. Emission Control)	AFA
C65, M65	M.T. 640G, 650G (exc. Exh. Emission Control)	AHH
C60-65	A.T. 540 w/Air Brake (exc. Exh. Emission Control)	AHJ

Model	Description	Code
S60	A.T. 540 (exc. Exh. Emission Control)	AFB
C60-65	A.T. 475 (exc. Exh. Emission Control)	AHL
T60-65	Manual Trans. (exc. Exh. Emission Control	AHU
T60-65	A.T. 540 (exc. Exh. Emission Control)	AHW
C60-65, S60, M65	Manual Trans. w/Air Brake & Exh. Emission Control	AHX
C65, M65	M.T. 640G, 650G w/Exh. Emission Control	AHY
C60-65, S60	A.T. 540 w/Air Brake & Exh. Emission Control	AHZ
T60-65	Manual Trans. w/Exh. Emission Control	ATB
T60-65	A.T. 540 w/Exh. Emission Control	ATC
C60-65, M65	Manual Trans. (exc. Air Brake & Exh. Emission Control)	AKK
C60-65	A.T. 540 (exc. Air Brake & Exh. Emission Control)	AKL
C60-65, S60, M65	Manual Trans. w/Exh. Emission Control (exc. Air Brake)	AKM
C60-65, S60	A.T. 540 w/Exh. Emission Control (exc. Air Brake)	AKR
427		
C65, M65	Manual Trans. (exc. Exh. Emission Control)	AKA
C65, M65	M.T. 640G, 650G (exc. Exh. Emission Control)	AKB
T65	Manual Trans. (exc. Exh. Emission Control)	AKC
T65	M.T. 640G, 650G (exc. Exh. Emission Control)	AKS
C65, M65	Manual Trans. w/Exh. Emission Control	AKD
C65, M65	M.T. 640G, 650G w/Exh. Emission Control	AKH
T65	Manual Trans. w/Exh. Emission Control	AKJ
T65	M.T. 640G, 650G w/Exh. Emission Control	AKX
1977—366		
C60-65, S60, M65	Manual Trans. (exc. Air Brakes & Exh. Emission Control)	AFH
C60-65, S60, M65	Manual Trans. w/Air Brakesa (exc. Exh. Emission Control)	AFD
C60-65, S60	A.T. 540 w/Air Brakes (exc. Exh. Emission Control)	AFF
C60-65, S60, M65	A.T. 540 (exc. Air Brakes & Exh. Emission Control)	AFJ, AKL
M65	A.T. 650 w/Air Brakes (exc. Exh. Emission Control)	AHH
C60-65, S60, M65	Manual Trans. w/Exh. Emission Control (exc. Air Brakes)	AKM
C60-65, S60	A.T. 540 w/Exh. Emission Control (exc. Air Brakes)	AKR
C60-65, S60, M65	Manual Trans. w/Air Brakes & Exh. Emission Control	AHX
C60-65, S60, M65	A.T. 540 w/Air Brakes & Exh. Emission Control	AHZ
C, M65	M.T. 640, 650G w/Air Brakes & Exh. Emission Control	AHY
C60-65, M65	Manual Trans. w/PTO and 4 BBL. Carb.	AFL
T60-65	Manual Trans. (exc. Exh. Emission Control)	AHU
T65	A.T. 540 (exc. Exhaust Emission Control)	AHW
T60-65	Manual Trans. w/Exh. Emission Control	ATB
T60-65	A.T. 540 w/Exh. Emission Control	ATC
427		
C65, M65	Manual Trans. (exc. Exh. Emission Control)	AKA
C65, M65	M.T. 640G, 650G (exc. Exh. Emission Control)	AKB
C65, M65	Manual Trans. w/Exh. Emission Control	AKD
C60-65, M65	M.T. 640G, 650G w/Exh. Emission Control & Air Brakes	AKH
C65	A.T. 540 (exc. Exh. Emission Control)	AJA, ALA
C65	A.T. 540 w/Exh. Emission Control	ALC
C65	Manual Trans. w/P.T.O. and 4 BBL Carb.	ALF
T65	Manual Trans. (exc. Exh. Emission Control)	AKC
T65	M.T. 640G, 650G (exc. Exh. Emission Control)	AKS
T60-65	Manual Trans. w/Exh. Emission Control	AKJ
T65	M.T. 640G, 650G w/Exh. Emission Control	AKX
T65	A.T. 540 (exc. Exh. Emission Control)	ALB
T65	A.T. 540 w/Exh. Emission Control	ALD

ending with JQ and a February 1969 production date. Refer to the Engine Series Number & Suffix tables on page 28 for 1969. The *JQ*, listed under Chevrolet, stands for 2-barrel carburetor, 3-speed automatic transmission—Turbo Hydra-Matic 400. Referring to the list on page 48, the only big-block Chevrolet with a two-barrel carb is the L-66. Now you have the key to what's inside the engine; cam specs, valve sizes, main-bolt numbers, and crankshaft material.

There are easier ways to find this information. Suppose you bought the car new and you remember exactly what you got. Or with luck, you kept the dealer's invoice. Good. But what are you going to use as a guide when you're looking for parts in a junkyard?

If an engine is still in the car—assuming it's the original engine—you have the VIN for reference. But what do you do when the engine is sitting on the ground and the car it was in is long gone? Fortunately, Chevy stamped the Engine Production Code and the sequence numbers on the block. Look on the small area of the block deck, or cylinder-head gasket surface that projects from the front of the right cylinder bank. What you need is the year and letter suffix, given by the Engine Production Code. Use these to identify the engine from the suffixes listed for that year on pages 28 and 29.

Unfortunately, the Engine Production Code will not always be there. Suppose you are in the junkyard looking at an engine or block. You have no clue as to what vehicle it came out of. So, you confidently stoop down and look expertly for the Engine Production Code. ''Gee, that's funny, there's nothing there.'' Blocks that were *decked*—their head-mating surfaces were milled—will be minus these stamped numbers; they were machined off. Now you need to rely on the *casting number.*

Casting Numbers—When Chevy designs a part, the engineering department assigns the part an *engineering number*. When the number is cast into the part, it becomes—you guessed it—a *casting number*. This helps to identify the parts in the factory for inventory and production scheduling until the part is put into the parts-distribution system. Then the parts department also gives the part a *Part*

Look for block casting number on top of bellhousing boss at at left rear of block. Fortunately, these casting numbers are large, making them easy to read if you can see them.

Number. This is an important distinction; *casting numbers are not part numbers!* Because a casting can be machined differently to make several different parts—or packaged differently—the parts department can't use the casting number for parts identification. And there is no listing of which casting is used for what part. All a parts man can use is the part number. For some reason several camshafts have their casting numbers listed in the parts books, but this is an exception.

Your situation is different. Because parts numbers don't appear on engine parts, but casting numbers do, use casting numbers to *help* identify parts. They will get you in the *ball park*. Use a part's physical description—as provided in this chapter—to complete its identification.

FINDING CASTING NUMBERS

Casting numbers won't do you much good if you don't know where to find them. This is not a problem if the part is sitting in the middle of a floor or bench. But a cylinder-head or -block casting number can be next to impossible to find on a grease-and-grime-covered engine in a cramped engine compartment. Here's where you'll find those casting numbers:

Cylinder Block—Look behind the left cylinder bank on top of the bell-housing boss. This number can be very difficult to see on an installed engine. But half the battle is won now that you know where to find it—and the numerals are big.

Heads—If you're lucky, the cylinder-head casting number will be in the area of the rocker arms, on top of a

With a little luck the head you are interested in will have its casting number here (arrow). There's also a casting number on the underside, but you'd have to remove the head from the block to see it. You'll only have to remove the rocker cover to see this one.

On this crankshaft, casting number is on the next-to-last counterweight/throw. Narrow line (arrow) is casting *flash*, identifying it as a nodular cast-iron crank. Wide flash line identifies a forged-steel crank.

center intake runner. If it's not there and the head is on the engine, you may have to remove it or use a mirror to see the casting number. For sure, the intake manifold has to come off. With the combustion chambers up and intake ports facing you, you should find the casting number under the far-left intake runner.

Cranks—The crankshaft casting number is between the *snout*—crankshaft nose—and the first counterweight or it's on the second-to-last counterweight.

Camshafts—Camshaft casting numbers are sometimes hard to read—sometimes they are illegible. You can either find the number immediately in front of the distributor-drive gear—that's where it is on *most* cams—or between the third and fourth bearing journal.

Once you've found the casting number on the part, refer to the appropriate chart to see which one you

CAMSHAFT-CASTING NUMBERS

Engine	Lifter	Year	Casting	Comments	Engine	Lifter	Year	Casting	Comments
366	H	66-68	4368	Gear-drive, truck	396 454	M	65-73	8911	
366	H	66-68	6350	Grooved rear journal					
366	H		6688		396 454	M	65-73	9604	
366	H	66-68	3856359						
366	H	69-74	3891516	Chain-drive	396	M	65-73	9180	Grooved rear journal. 1965-'66 ZL-1
366	H		3892342	Gear-drive, truck					
366	H		3906686	Standard rotation	454				
366 427 454	H	60-77	3935456	Chain-drive, truck	402 427 454	H	70-74	3545	Restamped 4365
396	H	65-66	4874	Grooved rear journal					
396	H	66-71	3944	High-lift, grooved rear journal	402 427	H	70-74	4365	Restamped 1516, 5465, 8345
402 454	H	71-77	3963545		454				
396 427	H	66-68	4864	Grooved rear journal	402 427 454	H	70-74	5465	Restamped 4365
396 427	H	66-68	4866		402 427 454	H	70-74	8345	Restamped 4365
396 427	H	66-68	3863144		427	M	67-74	4367	L-88 cam
396 427 454	H	66-74	4366	High-lift	427	H	69-77	0883	Chain-drive truck, no groove
396 427	H	67-74	4364		427	H	73	0888	Truck, no groove
					427	H	73	330885	Truck
396 454	M	65-73	5533	Gear-drive L-88. 1965-'66 versions have grooved rear journal	454	H	73-76	3041	

have. Zero in on parts identification by using the physical descriptions that follow.

CYLINDER BLOCKS

366 & 427 Truck—At 3.935 in., the 366-truck block has the smallest bore of any big-block. It shares crankshafts with the 396, 402, 427 and 427 truck. Here *truck* means medium or heavy trucks, not a pickup truck. Big truck engines displacing 427 cu. in. use a different casting, but otherwise differ from the 366 only in bore size. The 427 blocks have 4.251-in. cylinder bores.

The big difference between truck and passenger-car blocks results from differences in their piston *compression heights*—distance from center of the piston pin to top of the piston. Both the 366- and 427-truck blocks use four-ring pistons rather than standard passenger-car three-ring pistons. To make room for the fourth piston ring, piston compression height was increased on truck blocks. This required a 0.4-in. increase in block *deck height*—distance from main-

bearing-bore center to the block deck surfaces.

Passenger-car blocks have a 9.8-in. *deck height,* whereas truck blocks measure 10.2 in. Of course, as deck height increases, the distance between cylinder heads also increases. Consequently, both truck *tall* blocks require wider intake manifolds than their passenger-car cousins.

If you do run across a truck engine, don't discount it for passenger-car use altogether. With the exception of pushrods, pistons, manifolds and distributor housings; all truck-engine parts will interchange with passenger-car big-blocks. Just make sure the taller and wider engine will fit the engine compartment. If it won't, you can use it for parts. And if you have a high-performance application in mind, 427-truck blocks are great for building big-inch big-blocks by installing a *stroker* crankshaft.

Truck blocks can be identified by their wider intake valley and taller cylinder banks. Sometimes, but not always, they are cast with the word TRUCK in the timing-chain area or at

366- and 427-truck blocks have 0.4 in. more deck height to accommodate four-ring pistons. Standard block butted end-to-end with truck block shows difference. Photo by Bill Fisher.

BLOCK-CASTING NUMBERS

Casting	Bore (in.)	Deck Ht. (in.)	Engine	Year	Comments
345014	4.251	9.8	454	70-76	Chevy, GMC
359070	4.251	9.8	454	70-76	Chevy, GMC
3855061	4.094	9.8	396	65-67	
3855961	4.094	9.8	396	65-76	Chevy, truck. Also 350
	4.251		427		small-block
3855977	3.935	10.2	366	66-67	Chevy truck
3866961	4.094	9.8	396	66-69	Chevy, Chevy truck
3869942	4.251	9.8	427	66-72	Chevy
3902406	4.094	9.8	396	65-67	
3902466	4.094	9.8	396	65-67	
3904351	4.251	9.8	427	66-72	
3904354	3.935	10.2	366	66-76	Chevy truck, GMC
3916319	3.935	10.2	366	68	Chevy truck
3916321	4.251	9.8	427	66-69	
3916323	4.094	9.8	396	67-68	
3916328	4.094	9.8	396	65-67	
3918319	3.935	10.2	366	66-70	Chevy truck, GMC
3928319	3.935	10.2	366	69-73	GMC
3935439	4.251	9.8	427	66-69	
3935440	4.094	9.8	396	68	
3937726	4.094	9.8	396	68-73	
	4.251		427		
3955272	4.094	9.8	396	68-69	
3963512	4.251	9.8	454	70-76	Chevy, GMC
3965440	4.094	9.8	396	68-69	
3965449	4.094	9.8	396	68-72	
3965540	4.094	9.8	396	68	
3969854	4.094	9.8	396	68-72	Chevy, Chevy truck,
	4.126		402		GMC
6272176	3.935	10.2	366	68-76	Chevy, GMC
6272177	4.126	9.8	402	72-73	GMC
6272181	4.251	10.2	427	73-76	

Block-casting numbers are found on top of bellhousing boss at left.

Use both casting number and physical dimensions to identify a block. Rely on casting numbers only and you're asking for trouble as they aren't 100% reliable; physical features are.

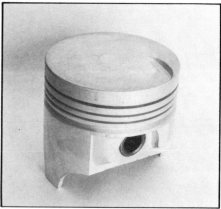

Count ring grooves; there are four. Truck-366 and -427 piston has three compression rings. Note also flat-top and valve-clearance notch.

the rear of the block. Inside, you'll *always* find four-bolt main-bearing caps. And you may find a gear-driven camshaft. The 366 engine used gear-drive cams in '66—'68; 427 truck engines in '68 only. Chain- and the reverse-rotating gear-driven cams will interchange, so don't be scared off by a gear-driven cam. Swap to a chain-drive setup if you don't want the noise that goes with a gear-drive camshaft. Just be aware that the cam must be changed with the drive components because of the difference in rotation.

When changing to a chain-drive cam, you can use the same distributor that was used with the gear-drive cam, just change to a 1958599 drive gear. It's needed because the of the reverse helix angle on the cam's

distributor-drive gear.

One thing to keep in mind: Trucks accumulate a lot of miles in a hurry, so most used truck blocks will require boring. This automatically means new pistons. Truck blocks, even by Chevy big-block standards, use thick cylinder walls. This means there's a lot of boring stock in a truck block—especially the 366 casting. The maximum practical overbore is 0.060 in. because of piston availability, however these blocks can be safely bored 0.125-in. over.

396—The most common big-block, the 396, has a 4.094-in. bore and 3.76-in. stroke. Many parts from this engine are found in other big-blocks, making parts scrounging a happy experience for the 396 owner. Cranks and rods from non-454 engines can be

used if rebalanced. In fact, 366- and 427-truck rods are the same ones used in the 396.

402—In 1970, Chevy began releasing "396s" with a 0.030-in. overbore. This increased the displacement of the 396 to 402 cu. in. Chevy didn't bother to change the medallions or stickers to reflect this change; they still said **396**. If you didn't know better, after looking at a car with a 402, you would think the engine was a 396. But a Chevy parts man knows there's a difference. They call '70 and '71 big-blocks *400's!*

Those of you who know your Chevys will realize there is another "400"—a small block. The parts man had to understand that the small-block 400 was offered *only* with a two-barrel carburetor and the big-block *only* came with a four barrel. This is the *only* differentiation made by the factory between these two, *completely different* engines.

Confusion ended with the 1972 models when Chevy identified the overbored 396 as a 402—both on the car and in the parts books. 1972 was also the last year for the 402. So, if you have a '69 or earlier car or pickup with a 396 nameplate, you really have a 396. 1970 and '71 cars have 402s, but with 396 nameplates. The only big-block offered after 1972 was the 454.

Not to worry now that you've found your engine is a 402, not a 396. Except for the 0.030-in.-larger bore and pistons, everything in the 402 is the same as a 396. The overbore gives the 402 a 4.124-inch bore. Crankshaft stroke remains at 3.76-in. Even the part number for all 396, "400 4-barrel" and 402 replacement crankshafts is the same, 3882841.

Chevy installed three different timing chains in the big-block. At left is a "roller-type" chain set installed in some truck big-blocks. Not a true roller, this chain is noisy. Center Morse-type chain-and-sprocket set came on all '65 and '66 big-blocks, and all performance applications from '67 on. This "silent-link" design is 3/4-in. wide. At right is narrower 5/8-in.-wide chain used in standard-duty engine from '67 on. Photo by Bill Fisher.

454 four-bolt block plumbed for an oil cooler. There's no doubt about whether this is a four-bolt block. The only way to be sure is to look at the caps.

If you have a 402, be alert for those easily made mistakes; especially by a parts man who might not know the difference between a '69 and a '70 "396." Play it safe when combing the junkyards, too. Measure the bore of *any big-block 396* with an Engine Production Code indicating that it is a '70 or later 396. *Be sure* of what you're dealing with.

427—One reason for the big-block's popularity is its high-performance potential—and one of the big reasons for this is the 427. Most 427s were offered in Corvettes—most of which bristled with performance options. Although it only had a four-year production run—1966—1969—the 427 was important in developing the high interest in the big-block.

To obtain the larger 4.251-in. bore, Chevrolet used a unique casting. The 427 crank is unique. Although stroke remained at 3.76 inches, balance between 396/402 and 427 cranks is different. With the exception of '69 LS-1 engines, the 427 was offered only in higher-than-base performance versions. As a result, you'll find typical 427s with high-lift camshafts, forged pistons and cranks and aluminum heads or blocks. These parts are covered in more detail later in this chapter.

454—Although 454 and 427 blocks share the same bore size, the castings are different. Also, the 454 has its own 4.00-in.-stroke crank, which is *externally balanced*. Some counter-weighting is done outside the crankcase, at the damper and flywheel, rather than with the crank counterweights only. *Internally balanced* engines are fully balanced by their crankshaft counterweights. Although 454 blocks will accept a 427 crank and rods, stuffing a 454 crank into a 396/427 block will require *clearancing* the block by grinding metal from the bottoms of the cylinders at their outboard edges. This is done to provide clearance for the connecting rods.

Any 454 block will interchange with any other 454 block. Your Chevy parts man can no longer supply you with anything except standard (STD), 0.020- and 0.030-in.-oversize pistons for your 454, but TRW has 0.060-over pistons available. Regardless, it's wise to check the availability of pistons *before* you bore.

High-Performance Blocks—With the exception of the two truck blocks, all *rats* came in high-performance trim. Block-wise, beefy four-bolt mains and provisions for mounting an oil cooler were available, depending on horsepower ratings. Two tapped holes above the oil-filter boss are for oil-cooler plumbing. These holes will be blocked with Allen plugs when an oil cooler is not used.

A block with oil-cooler provisions is not necessarily equipped with four-bolt main-bearing caps. Most high-performance, four-bolt blocks have **HI PER PASS** cast into the timing-chain area, but so do many standard-performance engines. A few have two tapped holes next to the front cam-bearing bore. When you find the tapped holes at the front and at the oil-filter boss, and **HI PER PASS** on the same block, chances are the block has four-bolt-mains. Of course, if the front cover is off, you'll be able to see the front cap between the crank nose and the oil pan. If the front cap has four bolts, you can be sure the other four do.

Aluminum—The RPO ZL-1 427 Corvette engine, available in 1969 and 1970, has an aluminum block and heads for an approximate 200-lb weight savings. It's highly unlikely you'll run across one of these in a junkyard—mainly because the ZL-1 option cost $3000 in 1969! Not too many were built and the survivors are prized collectors items. Aluminum replacement blocks are available from Chevy for a pretty penny; all have four-bolt mains. About the only people who can "afford" these blocks are professional racers.

Boring—If the standard bores are worn to the point of needing reboring, your best choice is an 0.030-in. overbore. Many machine shops have a policy of boring every block

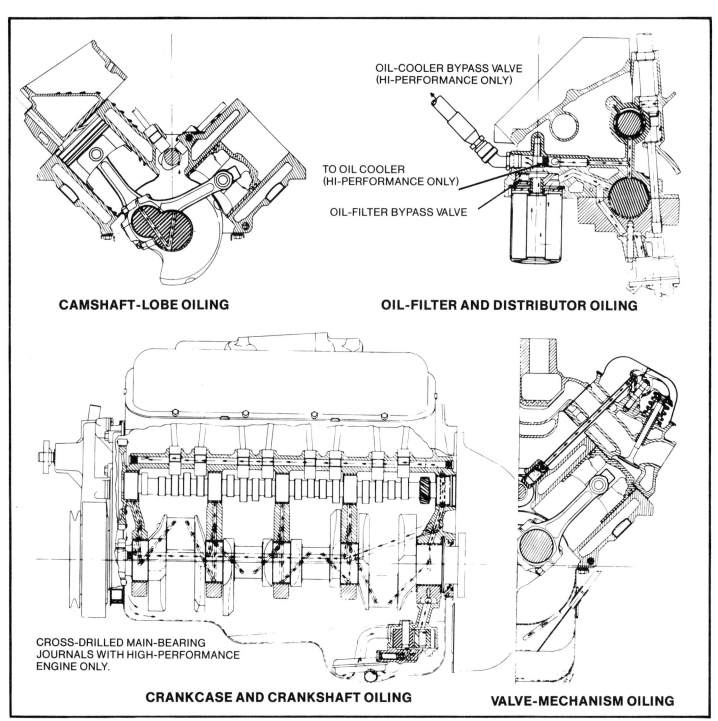

CAMSHAFT-LOBE OILING

OIL-COOLER BYPASS VALVE
(HI-PERFORMANCE ONLY)

TO OIL COOLER
(HI-PERFORMANCE ONLY)

OIL-FILTER BYPASS VALVE

OIL-FILTER AND DISTRIBUTOR OILING

CROSS-DRILLED MAIN-BEARING
JOURNALS WITH HIGH-PERFORMANCE
ENGINE ONLY.

CRANKCASE AND CRANKSHAFT OILING

VALVE-MECHANISM OILING

Drawings illustrate why big-block Chevys have a well-earned reputation for excellent lubrication. Note hollow pushrods and how cylinder walls receive oil from grooved connecting-rod cap. Early grooved rear cam-bearing journal and bearing are shown. Note also plumbing for oil cooler. Drawing courtesy of Chevrolet.

0.030-in. oversize to save time. Why? Although most bores *clean up* at 0.010-in., it's not uncommon that at least one bore requires 0.030-in. over to clean up. Starting with a 0.030-in. overbore eliminates the need for boring a block twice and losing time.

Another limiting factor is piston availability. Chevy offers only STD and 0.030-in.-over pistons for 396 through 427 engines. There are 0.020-in.-over pistons available for

the 454. This is a result of Chevy's consolidation efforts on out-of-production engines. All is not lost. TRW offers 0.030-, 0.040- and 0.060-in. oversize pistons, all forged.

If more than a 0.030-in. overbore is required to clean up the bores of a 396 or 402, it's possible to go all the way to the 427/454's 4.251-in. bore. You can use standard 427 pistons. *Don't install 454 pistons.* They have 0.130-in. less compression height.

Be forewarned: *Boring a 396 or 402 to 4.251 in. is a risky proposition—it may not work.* The problem is you won't find out until after the engine is back in operation and a cylinder wall fails. A safe and sane overbore limit for the 454 is 0.060 in. Go larger and you'll have trouble getting pistons with a low-enough compression ratio. Plus, limiting overbore to 0.060 in. will leave enough cylinder-wall thickness for trouble-free *street service.*

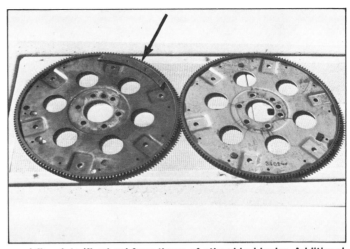

Counterbalance weights (arrows) distinguish a 454 crankshaft damper and flexplate/flywheel from those of other big-blocks. Additional 0.24-in. stroke of the 454 crank requires the use of external counterbalancing. Use these parts incorrectly and the engine will shake off of its mounts.

Truck blocks have thick cylinder walls and can be enlarged 0.125 in.; 366 blocks are especially thick.

For the record, the aluminum ZL-1 can take only a 0.030-in. overbore. After that it's time for new *liners, or sleeves.*

Racing Block—There's also a 4.440-in.-bore aluminum-alloy block that is sold for racing only. The cylinders do not have separate *liners,* or sleeves—the pistons run directly on specially treated aluminum bores. Chevrolet suggests a 0.060-in. overbore. With a 4.00-in.-stroke crankshaft, this gives 510 CID!

CRANKSHAFTS

Swapping big-block crankshafts is simple. With the exception of the 454, all big-block crankshafts will interchange. For starters, all have the same diameter journals. Connecting-rod journals are 2.200 in.; main journals are a hefty 2.7495 in.—except the rear, or number-5, journal. It is slightly smaller at 2.7488 in.

454—Stroke and balance makes the 454 crank different from others. All 366, 396/402 and 427 cranks have a 3.76-in. stroke and are internally balanced; 454 cranks have an even 4.00-in. stroke and are externally balanced.

External balancing is done at the 454's crank damper and flywheel. Counterbalance weights added to the damper and flywheel compensate for the out-of-balance resulting from the stroke increase. Therefore a 454 crank—regardless of the block it's installed in—must have a 454 damper and flywheel or flexplate bolted to it to be in balance.

You may have heard of internally balancing a 454 crank by adding *heavy metal* to its counterweights. Heavy metal is nothing more than a high specific-weight metal, such as lead. The procedure is possible, but it's a job only for an expert crank man—and it's quite expensive. Your best bet, by far, is using a counterbalanced flywheel and damper with a 454 crank.

Another consideration is fitting a 454 crank inside a 366, 396/402 or 427 block. The crank will bolt in, but the rods will interfere with the block at the bottom of some cylinders. You'll have to do some checking and grinding to make the combination work.

An exception is the late-427 block. This block uses the same casting as early 454s, so there's no clearance problem. If you'd like the 4-in.-stroke 454 crank in your block, don't forget to consider the time and cost of grinding clearances. There is also the cost of the crank, damper and flywheel. With all these points in mind, leave a 454 crank in a 454 block.

On the other hand, the internally

Installing a 454 crank in a non-454 block may require *clearancing* the block at the bottom of the cylinders for rod clearance. Trial fitting the crank, rods and pistons is required to determine exactly how much and where this clearancing must be done. Unless you are building a racing engine, stay away from this swap. It can be costly and time consuming. Photo by Bill Fisher.

CRANKSHAFT-CASTING NUMBERS

Engine	Casting	Engine	Casting
366, 396,	6223	454	3521
427, 427T	7115	(4.00-in. stroke)	7416
(3.76-in. stroke)	3863144		353039
	3874874		3963523
	3882842		3963524
	3882847		3967463
	3882848		3975945
	3882849		
	3887114		
	3904815		
	3904816		
	3942411		

CRANKSHAFT-BEARING JOURNALS (in.)

Rod Journal Diameter	Main Journal Diameter #1-4	Main Journal Diameter #5	Thrust Faces Overall Width
2.200	2.7495	2.7488	1.8170

Although all identification numbers are referred to as "casting numbers," forged cranks are included. Flash-line width distinguishes a cast-iron crankshaft from a forged-steel one. If you are still in doubt, tap a counterweight with a hammer. A forged crank will "sing." A cast crank will respond with a "thud."

balanced 366—427 crank needs no additional counterweighting on the crank *outside* of the engine. In other words, the rotating masses of the 366—427 engines are balanced entirely *inside* the crankcase by the crankshaft counterweights. Harmonic dampers and flywheels for these engines are *zero balanced*—the damper and flywheel are not counterweighted. This means you can use a 427 flywheel on a 396, or a 402 harmonic damper on a 427.

396/402 & 427—As similar as they are, there is a difference in counterweighting between 396/402 and 427 crankshafts. The 427 crank has heavier counterweights than the 396 to compensate for the heavier 427 pistons.

You can tell a 396/402 crank from a 427 by the width of the third counterweight. Count from the front of the crank to the third counterweight. The third 396 counterweight is 7/16-in. wide; the 427 measures 7/8 in. The two cranks are interchangable, but you'll have to rebalance the engine.

FORGED STEEL VS. NODULAR CAST IRON

Both *forged-steel* and *nodular-cast-iron* cranks have been used in the big-block Chevy. Forged steel is much stronger than nodular cast iron, and is usually found only in racing engines. Big-block Chevys are an exception to this rule. Only the lowest-horsepower versions use nodular-iron cranks. These engines are the two-barrel 265-HP 396, available only in 1969, the 325-HP 396 in 1968 and '69, 240/210-HP, 300-HP and 330-HP 402s, and the 335-HP 427. Through 1972, all 454s had forged-steel cranks. 1973 and later 454s use nodular-iron cranks.

If your engine came with a nodular-iron crank there is no need to swap for the forged-steel variety. As long as you rebuild only to stock performance specifications, the nodular crank is strong enough.

Distinguishing a forged crank from a cast one is easy: Look for the parting line that runs the length of the crank on the as-cast or as-forged unmachined surfaces. The nodular-cast-iron crank has a thin parting line, or *flash,* where the mold halves separated. Forged cranks have a much wider parting line at the die halves. Where the casting might have a 1/32-in.-wide parting line, the forging line will measure about 3/4 in. Additionally, the unmachined surfaces of a forged-steel crank are much smoother than those of a cast-iron version.

Another method to tell a forging from a casting is *by the sound it makes.* When tapped by a metal object, such as a hammer, a forged crank will "sing." You'll hear nothing but a thud as you strike a cast crankshaft. This is because of the ability of cast iron to damp high-frequency vibrations—an advantage of a cast-iron crankshaft.

HI-PER Crankshafts—As there are high-performance blocks, there are also heavy-duty/high-performance cranks. Those 396 engines, rated at 375 HP or more, and 427s or 454s of more than 425 HP, have their cranks *Tuftrided*—a chemical, heat-treating process used to harden bearing and journal surfaces. Tuftriding leaves a dull-gray finish.

Many big-block crankshafts, including those in 366- and 427-truck engines, also have *cross-drilled* main-bearing journals. Cross-drilling refers to extra oil passages drilled in a bearing journal to ensure that the bearings and bearing journals are supplied with oil through 360° of crankshaft rotation.

All Tuftrided cranks are cross-drilled, but not all cross-drilled cranks are Tuftrided. Neither of these heavy-duty features are needed for street driving, but were offered with big-blocks intended for high-performance

Wide flashing on this crank (arrow) indicates it is forged steel. Cross-drilled bearing journals ensure a constant oil supply to the rod journals. Although this cross-drilling was done by a machine shop, cross-drilled cranks were standard in truck and some high-performance big-blocks.

High-performance rod at left and standard version at right. Bump on beam web near pin-end (arrow) identifies a high-performance rod. A close look reveals that the high-performance rod is beefier, particularly at the big end.

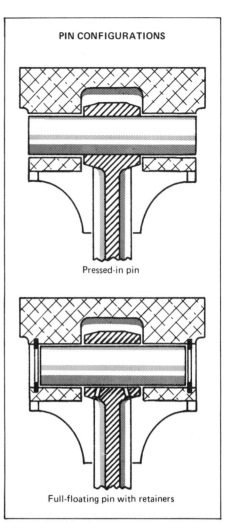

PIN CONFIGURATIONS

Pressed-in pin

Full-floating pin with retainers

Not all big-blocks use pressed-in piston pins. L-88 and ZL-1 engines came with floating pins, bottom. Floating pins are held in place with retaining rings at each end.

or heavy-duty use, even though the stock bottom end is plenty tough.

Make sure any crank you buy is free of cracks, or scratches or scored journals. Refer to pages 75—78 for details on inspecting a crankshaft. Give the flywheel flange and bearing-journal radii a thorough examination for cracks: these are common breaking points. To be positive, have the crankshaft *Magnafluxed,* page 78.

CONNECTING RODS

Another easy-to-swap part is the big-block connecting rod. If you look at major design parameters only, such as center-to-center length, sorting out big-block rods is easy—there's only one type to worry about. High-performance rods and bolt types create a few variations, but when it comes to a standard rebuild there aren't any complications.

The big-block rod has a 6.135-in. *center-to-center length*—distance from the center of the *big-end*—crank-end bearing bore—to the center of the piston-pin bore. This measurement is the same for all big-block connecting rods, regardless of stroke. Stroke is changed only by altering crankshaft throw, not by changing rod length. And furthermore, all rods have 2.325-in.-diameter big-ends.

A standard-duty rod, 3933174, may have a dab of orange paint on its beam. Otherwise, standard-duty rods can be identified by their 3/8-in.-diameter bolts. These bolts have *knurled shanks.* The bolt's

shank—the length between the bolt head and the threads—has a raised cross-hatch pattern. The knurl causes each rod bolt to fit snugly in its rod and accurately positions the bearing cap on the rod.

High-Performance Rods—Variations surface when you get into high-performance options. Later high-performance engines, including the L-88 and ZL-1, came with beefier rods. The rod-beam web has a small bump immediately below the pin end. See above photo: A blue paint dab *may* identify this connecting rod. These rods were fitted with 7/16-in. bolts with ground shanks—they are not knurled.

Another difference: Some high-performance connecting rods were equipped for use with *floating* piston pins—the pin is free to move both in the piston and the rod's pin bore. Unlike the standard big-block rod, the small-end bore provides 0.0001—0.0008-in. clearance to the pin. This allows the pin to rotate in the rod as well as the piston. Longitudinal pin movement is restricted by retaining rings installed in grooves in each end of the piston-pin bore. Standard pressed-in pins have about a 0.001-in. interference fit in the rod for retention.

Yet another high-performance rod is the one used in the LS-6. It uses the same forging as the L-88 rod, so it has the identifying bump. The LS-6 rod also uses 7/16-in. bolts, but with knurled shanks. This connecting rod

is stronger than the standard rod with its 3/8-in.-bolts, but not quite as muscular as the late L-88 rod.

Some machinists say you will find a pink color code on some high-performance rods; so, if you have pink rods, congratulations!

Unless you stumble across a set of high-performance rods at a really low price, stay with the standard rods; they are strong enough for street operation. As evidence of this, you'll find them in 366 and 427 truck engines. If you need more convincing, check Chevy's prices: *$51 per rod* in 1983!

Piston Pins—All big-block Chevys use the same 0.9897-in.-diameter piston pin. This is the case for both pressed-in and floating pins. And in case you were wondering, pressed-in pins are best for all applications. Floating pins offer no advantages other than ease of assembly.

Expansion slots are above each skirt just below the oil-ring groove. Cast pistons are also identified by steel struts cast into pin bosses. In this photo you can see the strut on each side of the far pin boss. Note also pin-oiling holes.

TRW forged pistons have a grooved pattern machined into skirts to retain oil for piston/cylinder-wall lubrication. Atop piston are both part number and size. This domed piston is for a 0.030-in.-overbore. What doesn't show is *barrel* shape machined into skirt.

PISTONS

Two basic types of pistons are used in the big-block Chevy: *forged* and *cast*. Although both are aluminum, different methods are used to manufacture the pistons. Forged pistons—beginning as a solid billet—are formed under extreme pressure in a die. A cast piston is made from molten aluminum poured into a mold.

Forged pistons are stronger and more resistant to high temperatures than cast pistons, making them better for high-performance use. However, cast pistons are the first choice for street engines because they can be closely fitted in their cylinder bores for quiet operation and good oil control. More importantly, cast pistons wear longer than than their forged counterparts because of superior oil-retention. Simply put, they will go more miles because they are better lubricated.

All '65—'67 396s and 402s are equipped with with forged pistons. In 1968, cast pistons were introduced on the standard-performance engines: all 396s and 402s under 375 HP, and all 427 and 454 engines under 400 HP. The last forged pistons were offered in 1971 on 454s. Engines that got forged pistons include: 375- and 425-HP 396s and 402s, 425- and 435-HP 427s and 425-, 450-, 460-, 465-HP 454s. The rule of thumb is: mechanical cam, forged pistons; hydraulic cam, cast pistons. This rule does not apply to early 396 and 427 big-blocks.

Stay with cast pistons if they were originally in the engine and it won't be used for racing. True, forged pistons are stronger, but it's rare that a street engine is ever exposed to the high temperatures and shock loads that require their use. A cast piston needs only 0.0007—0.0015-in.-bore clearance. This is because of its low thermal expansion; a cast piston's diameter increases little as it heats up to operating temperature. Steel *struts* are cast integrally into the piston for dimensional control.

On the other hand, forged pistons require *a minimum* of 0.002-in. clearance; most need almost 0.004-in. That's why forged pistons are noisier on initial startup than their tighter-fitting cast-aluminum counterparts. And don't forget the superior wear qualities of a cast piston—a major consideration for a street engine.

Piston Identification—To identify a cast piston, look for a slot at the back of the oil-ring groove above each skirt. This, along with the cast-in struts, is part of a cast piston's thermal-expansion control. Forged pistons have no slots, only a *barrel-shape* skirt for expansion control.

Another identifying feature of a cast piston is its underside, particularly around the pin boss. It is made up of many geometric shapes, compared to the relatively smooth and simple underside of a forged piston.

Floating Pins—Look for the retaining-ring grooves at both ends of a forged piston's pin-bore. They'll be there if floating piston pins are used. Cast-aluminum pistons do not have floating pins, nor is there room to install retainers for floating pins, so don't try. Forged pistons can be modi-

fied to accept retaining-ring grooves if you want. You can also widen existing grooves—from the early 0.050-in.-thick retaining rings—so the later 0.072-in. Spirolox retainers can be installed. Neither modification is needed for street use, but the wider retainer is good insurance against breakage.

Piston Domes—There's a direct correlation between big-block horsepower and the "rise and fall" of the domed piston. Small domes were offered on the earliest big-blocks, but soon rose to towering heights in search of higher and higher compression ratios.

With the advent of emission regulations and poor gasoline, piston domes shrank to well below their starting point until the piston tops became little more than shrunken wells. As I explain later, dome shape is directly related to combustion-chamber shape. So there's really no "good" or "bad" piston-dome shape. The smoother the dome, the better the flame travel, or *propagation*. A high-dome, or popup, acts as a barrier to the flame as it travels from one side of the combustion chamber—beginning at the sparkplug—to the opposite side. Regardless, combustion-chamber shape, and valve placement and lift dictate dome shape for a given compression ratio.

Compression Height—One piston dimension you need to concern yourself with is *compression height*—distance from the piston-pin-bore center to the flat surface at the top of the piston periphery. Changing compression height allowed Chevy engineers to use the same-length connecting rod in *all* big-blocks, even with different-stroke crankshafts. The pin was raised for longer-strokes.

If piston compression height is indiscriminately changed, pistons will either hit the cylinder heads or be too far down the bores, resulting in severe engine damage or very-low compression. Many different pistons were used in the big-block Chevy—if you consider the multitude of piston-top configurations.

Standard Pistons—The first is the 396—427 height of 1.770 in. Shorter by 0.130-in. is the standard 454 piston at 1.640 in. Last are the tall 2.170-in. 366- and 2.160-in. 427-truck pistons. The 454 engine requires a piston with less compression height because of its long stroke. The two truck engines have the short-stroke crankshaft of the 396—427 engines, but four-ring pistons required increased compression height *and* cylinder-block-deck height. All 402s have a compression height of 1.770 in. Complexity appears as soon as you begin talking high-performance.

High-Performance Pistons—Except for the 1968 *heavy-duty* 396, 396-performance models use a 1.765-in. compression height. The heavy-duty 396 uses a lower, 1.760-in. compression height. Almost all 427s were high-performance engines. In four model years Chevy managed to use pistons with two different compression heights in the high-performance 427—1.765 and 1.760 in. High-performance 454s have a 1.645-in. compression height, 0.005-in. more than the standard-454 piston.

In summary, big-block Chevy piston compression heights are:

Standard	Compression Ht. (in.)
396—427	1.770
454	1.640
366 Truck	2.170
427 Truck	2.160

Performance	Compression Ht. (in.)
396*	1.760
396	1.765
427*	1.760
427	1.765
454	1.645

*1968 heavy-duty only

Selecting Pistons—It pays to be careful when choosing pistons. Your best bets by far are the high-quality Chevy or TRW replacements. However, if your engine is one of the higher-compression models and you just

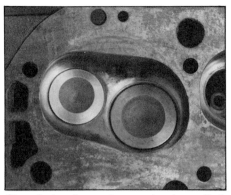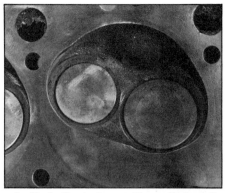

Closed-chamber and open-chamber cylinder heads. Photo illustrates problem with installing open-chamber heads in place of their closed-chamber counterparts. Although open chamber is larger, chamber is much tighter to the exhaust valve than with the closed chamber. This will cause a close-chamber-piston dome to strike the head in the area of exhaust valve.

can't stop it from detonating and dieseling, change to lower-compression pistons and heads or a combination of the two.

Refer to the cylinder-head chart on page 40 for an existing piston/cylinder-head combination that may provide you with the desired compression ratio. Read the cylinder-head section to see how bolting on different cylinder heads can give you the compression ratio you're after. Incorrect cylinder-head swapping can be disastrous. So do your homework and double-check every specification before installing different heads or pistons! If you can't make a matchup using Chevy parts, TRW makes lower-compression *Turbo* pistons for the big-block.

Whatever combination of pistons and heads you try, attempt to reach a compression ratio of not more than 9:1. It's the best compromise to maximize power and minimize detonation with today's gasoline. A sharply tuned engine running on premium fuel can get by with as high as 10.25:1 compression, but it may require a degree or two of spark retard—and probably water injection, too. This depends on numerous factors. Counting on the future availability of premium fuel is too great a risk, so you'll want to do something about an overly high compression ratio.

Water injection will control a lot of compression-related problems—namely detonation—while retaining the power advantage of 9:1 and higher compression. Read about water injection on page 93. If you decide to go with water injection, you can retain the stock pistons and heads—if they give less than 10.5:1 compression.

Another detonation-controlling device retards ignition timing. For manufacturers of these spark-control devices turn also to page 93.

Another piston-related factor is *flame-front* travel. Flame front is the term for the leading edge of the rapidly expanding mass of combustion gas as it burns the compressed air/fuel mixture, starting at the sparkplug and traveling to the outer walls of the combustion chamber. The flame front travels across the combustion chamber in a few thousandths of a second. The shape of the combustion chamber is carefully designed to control flame-front travel so the fuel charge *burns* smoothly and completely to pressurize the combustion chamber gradually and force the piston downward.

A piston-dome shape that doesn't match the cylinder-head combustion chamber may block flame-front travel. Abnormal, instantaneous combustion, or *detonation,* may result with gasoline of an inadequate octane rating. The results: engine damage, wasted fuel and lost power.

Chevy spent considerable time and money on piston-dome development, so, what's the problem? None, except many of these pistons were designed when premium fuel was plentiful and rated around 100 octane. What was acceptable then may not be now. So if you have a chance to change to lower-compression pistons, consider a flat-top design that does not obstruct flame-front travel. Your engine shouldn't have detonation problems if you do.

Just how important is good, even flame-front travel? That depends on your engine, how you drive, available gas and a continuing list of variables.

CYLINDER-HEAD CASTING NUMBERS

Casting Number	Combustion-Chamber Volume (cc)	Valve Size (in.) Intake/exhaust	Engine	Year	Comments
336768		2/06/1.72	427	73-76	Truck
336781	110.0	2.06/1.72	454	74	
343783		2.06/1.72	454	70-76	Chevy, GMC
346236	112.0	2.06/1.72	454	75-76	Truck, very small port
352625		2.06/1.72	454	70-76	Chevy, GMC
353049	110.0	2.06/1.72	454	73	
3846206	96.4	2.06/1.72	396	65-68	
3856206	96.4	2.06/1.72	396, 402, 427	65-73	Chevy, GMC, Chevy Truck
3856208	106.9	2.19/1.72	396, 427	65-67	425HP 427
3856213		2.06/1.72	366, 427T	65-76	Chevy Truck, GMC
3856260			396	68	
3872702	96.4	2.06/1.72	396, 427	65-66	325HP 396, 390HP 427
3873702	96.4	2.06/1.72	396	66	325/360HP
3873858	106.9	2.19/1.72	396, 427	65-66	375/425 HP 396, 425HP 427
3904390	96.4	2.06/1.72	396, 427	65-66	325/360HP 396, 390HP 427
3904391	106.9	2.19/1.72	396, 427	65-66	375/425HP 396, 425HP 427
3904392	103.3	2.19/1.72	427	67	Aluminum 430/435HP
3909802	96.4	2.06/1.72	396	65-66	375/425HP 65 only, 375HP 66 only
3917215	96.4	2.06/1.72	396, 427	65-68	325HP 396, 390HP 427
3917219	96.4	2.06/1.72	366, 427	66-73	
3919839	104.9	2.19/1.72	396, 427	68-69	High-Perf. Large-Port
3919840	104.9	2.19/1.72	396, 427	65-69	High-Perf. Large-Port 375HP 396 425/435HP 427
3919842	103.3	2.19/1.72	396, 427	68-69	Aluminum, Large-Port 375HP 396 435HP 427
3931063	96.4	2.06/1.72	396, 427	65-69	325/350/385 & 390HP 396
3933148	109.14	2.06/1.72	396	69	265HP 396
3946074	114.8	2.19/1.88	454, 427	69-71	Aluminum '69 430HP 427, '70 465HP 454, '71 425HP 454
3964280	96.4	2.06/1.72	396, 454	65-72	325HP
3964290	96.4	2.06/1.72	396, 454	65-76	Chevy, GMC 345, 360, 390HP 454
3964291	106.9	2.19/1.88	396, 402 427, 454	65-70	High-Perf. 425, 435HP 427 in '69
3964292	106.9	2.19/1.88	454	70	High-Perf. 450HP
3964380			396	69-72	
3965198		2.06/1.72	396, 402	68-71	Chevy, Chevy Truck, GMC
3975950		2.06/1.72	396, 402	70	Chevy Truck, GMC
3986135		2.06/1.72	427	69-73	GMC
3993820	105.0	2.06/1.72	402, 454	70-72	Chevy, GMC
3994026	103.1	2.19/1.88	454	71	425HP
3999241	105.0	2.06/1.72	402, 454	71-72	365HP 454 in '71
6258723	103.1	2.19/1.88	454	71	425HP
6272290		2.06/1.72	402	72-73	GMC
6272990	103.1	2.19/1.88	454	71	425HP

Combustion-chamber volume is only listed where good information was available. Always double-check a part's identity through its physical features.

CYLINDER HEADS

Open & Closed Chamber—A good way to describe big-block heads is to categorize them as *open or closed chamber*. Closed-chamber cylinder heads have small combustion-chamber volumes for high compression ratios, with even mildly domed pistons. Open-chamber designs are larger and provide lower compression ratios, so require a much larger piston dome to achieve the same compression as a closed-chamber head.

Consult the cylinder-head table, read about anti-detonation strategy, page 93, then make your decision.

Before you do any cylinder-head interchanging, make sure the pistons are compatible with the heads. Use casting numbers *and* physical features to identify cylinder heads. Above all, don't assume a cylinder head will work simply because its combustion-chamber is bigger.

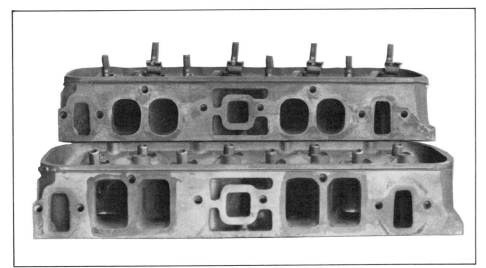

454 low-HP head, top, versus LS-6 454 at bottom. Low-performance cast-iron heads have oval intake ports; high-performance ports are rectangular.

Small is the word for late-model-truck intake ports, above. Although truck ports are *oval*, there is a difference amongst oval ports. Unlike the intakes, exhaust ports on these heads are the same size as other big-block oval-port heads. Small, oval-port heads are great for low-rpm torque; perfect for trailer towing.

Closed Chamber—Closed chambers were the rule up to 1970, except for the high-performance L-88 and ZL-1 engines. You can spot a closed-chamber head simply by looking at the combustion-chamber shape. The chamber walls closely surround the valves and the sparkplug—see photos, page 42. The effect is a bath-tub shape with the valves in the bottom and the sparkplug at one side.

Open Chamber—Emission regulations resulted in the open-chamber heads. By 1970 the big-block was considered too "dirty" and couldn't operate efficiently on low-octane fuels with the high-compression, closed-chamber head. So Chevy redesigned the combustion chambers.

Opening up the area around the sparkplug reduced compression and, more importantly, removed most of the chamber's *quench*. Quench refers to a "tight squeeze" designed into the combustion chamber by shaping the piston and cylinder heads so they are very close to each other at TDC. This quench area keeps the air/fuel charge cooler to avoid detonation.

Unfortunately, the quench area also forms pockets where unburned combustion gasses collect, causing unwanted emissions. By eliminating the quench area around the plug and valves, the big-block was made to run cleaner. Interestingly, the open-chamber head developed more horsepower, even with lower compression. Opening the chambers *unshrouded* the valves, resulting in improved breathing.

Still more chamber-volume was added to the 1971 heads of the lower-performance engines. The area around the valves, opposite the sparkplug, was hollowed out. This removed another quench area and lowered compression still further. These heads *flow* combustion gases even better than previous open-chamber designs. As a result they regained some of the horsepower lost from lowered compression.

Another very important point concerning open- and closed-chamber heads involves modifying closed-chamber heads—reworking them into an open-chamber configuration. You may have heard this is possible, but 'taint so. You can remove a little metal from the sparkplug side of the chamber, but you'll soon break through into a water passage. If you have closed-chamber heads and want open chambers, you'll have to get a pair of open-chamber heads.

Small & Large Ports—Another major head-identifying characteristic is port size. Passages inside a head that connect intake and exhaust valves with their respective manifolds are called *ports*. Cast-iron heads with 2.06-in. intake valves all have oval-shaped intake ports. Rectangular intake ports in cast-iron heads are larger and have 2.19-in. intake valves. This is the key: Intake-port size in *cast-iron heads* is keyed to intake-port shape. Large intake ports have a rectangular cross section; small ports are oval. All exhaust ports in cast-iron heads are square with rounded corners. And they are the same size.

Even though cast-iron-head port shapes follow the large/small-port rule faithfully, actual intake-port

Cast-iron, open-chamber, large-port head is marked well. The HI-PERF designation is large and hard to miss. Chevy means what they say—this head was designed for racing. Running these heads on the street has been done, but low-speed performance is poor.

Just as high-performance heads are marked HI-PERF, truck heads are marked TRUCK. Expect heads with this designation to have very small intake ports for low-end power. There is some bad news: Don't expect sparkling top-end performance.

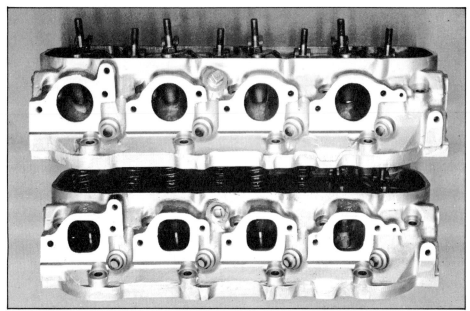

Aluminum heads: Classic examples of open- and closed-chamber heads, at top is ZL-1 and 2nd-design L-88; bottom head is 1st-design L-88. Round exhaust ports of 2nd-design L-88 heads are a big-block oddity. They didn't work too well and later heads went to C-shaped exhaust ports.

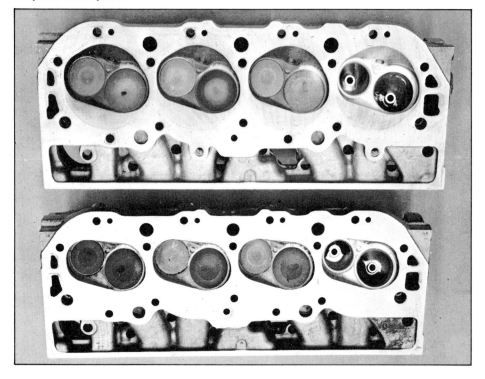

dimensions are not all equal. There's a big difference between large and small ports, but some small-port-head ports are smaller than in others.

For instance, small intake ports found in late truck heads: These ports are quite small for good low-rpm throttle response and torque required of a large-truck engine. You can use these heads on passenger cars and pickups, but bolting them onto a performance big-block will limit its high-rpm power.

Aluminum Heads—As with aluminum blocks, you won't find many aluminum heads laying around. Plenty of the high-performance options, including the L-88, L-89, late LS-6 and ZL-1, feature aluminum heads. They give about a 70-lb weight savings. An oddity found only on aluminum heads fitted to aluminum blocks—ZL-1s—is extra retention. Two studs on each head extend through intake ports and into the lifter valley to prevent blown head gaskets.

In case you find a pair of aluminum heads, check the intake ports for these extra holes. If you don't find them, go ahead and run the heads as is. If you find the two holes, you've got your hands on a pair of valuable ZL-1 heads. You can sell them at the next Chevy swap meet for big money. If you really want them on your iron big-block, you'll have to tap the extra holes above the ports and install threaded plugs. Don't do this without giving it a second thought: I think the heads are worth more with untapped stud holes.

Because aluminum heads were offered only as a high-performance option, they all have large ports. Those that first reached the market were on the first-design 1967 L-88 big-block. These closed-chamber heads were fitted with 2.19-in. intake valves and 1.84-in. exhausts. When the L-88 was redesigned for the ZL-1's introduction in 1968, the lightweight heads got open chambers and bigger exhaust valves: 1.88-in. Exhaust-port shapes were also changed; they went from rectangular to circular.

After 1971, cast-iron-head development was relatively tame compared to the changes made to aluminum heads during the same period. This is because iron heads were installed on passenger cars and truck engines that were already well developed—except for minor changes to reduce emissions. However, aluminum-head development continued because of the racers' demand for more horsepower—and Chevrolet engineering accommodated them.

So, an aluminum-head change took place in 1979. The round exhaust ports of the second L-88 design were changed to a C configuration. The port looks like a C with the open side facing down. A small protuberance inside the exhaust port enhances the C shape.

The latest aluminum head features a D-shaped exhaust port. In addition, the intake-port floor was raised to give the cylinder-head porters more material to work with. None of the C- or D-port heads were installed on the production line. They are strictly over-the-counter racing heads.

VALVE SIZES

Intake—To repeat, Chevy started with 2.06-in. intake valves in the big-block. They were enlarged to 2.19-in. for performance applications. There is a larger intake valve for the big-block Chevy: 2.30-in. This valve was never available in a cylinder head from Chevrolet, past or present. Available from Chevy dealers as a service part, the 2.30-in. valve was intended only for racing engines. But 2.30-in. valves weren't accepted. They proved too big—even for the big-block.

Exhaust—Like intakes, there are three exhaust valves. Unlike intakes, all have been used in production. All cast-iron, closed-chamber heads use 1.72-in.-diameter exhaust valves. Open-chamber iron heads use larger 1.88-in. exhausts, as do open-chamber aluminum heads. Closed-chamber aluminum heads use an in-between, 1.84-in. exhaust valve.

Which Size is Best?—For racing, the biggest valve-and-port combination is *usually* desirable. Not so for street use. Big ports and valves mean poor driveability. However, using the larger valves—2.19-in. intakes, and especially 1.88-in. exhausts—will not hurt low-rpm torque. But 2.19-in. intakes will give more power at higher rpm. Consequently, they are good for all-around power. However, for street driving, shy away from large-port heads.

Smaller ports keep air/fuel *mixture velocities* high—decreasing the problems of fuel condensing on the intake-port walls. Big ports and big valves make for a real dog in around-town driving; chugging away from each stoplight, then lugging up through the gears until cruising rpm is reached. From 3000 rpm up, or whatever, an engine with big-port heads will deliver more than adequate horsepower. By contrast, small-port heads, even in combination with large valves, will *pull* well from idle and also perform well on the highway.

Valve Guides—Chevy has a little surprise in the valve-guide department. Unlike small-block heads that use integral guides, big-block guides are pressed-in. You might know this type of guide as a *false guide*. The factory installs semi-finished guides, reams the guides and then machines the valve seats so they'll be concentric with their valve-guide bores.

Another surprise is that Chevy doesn't sell replacement guides, only replacement heads! Not to worry; worn guides can be renewed by install-

Plug type can date a big-block cylinder head. Even though all use 14mm threads, they differ in hex size, sealing and/or reach. Bottom to top; pre-'70 iron heads use 13/16-in.-hex plug with gasket, 3/4-in. reach, and partial threads. Aluminum heads also use the larger 13/16-in. plug with gasket, but with full threads. Small 5/8-in. plug with tapered seat is used in all cast-iron heads after 1969. Reach is about 1/2 in.

HEAD GAMES

Confused?—It's easy to have all the big-block head offerings blend into a hazy jumble. To reduce some of the confusion, here are some simple rules.

First, the cast-iron heads:
1. Oval intake ports mean small ports.
2. Rectangular intake ports mean large ports.
3. Small ports mean 2.06-in. intake valves.
4. 2.06-in. intakes also mean 1.72-in. exhausts.
5. Large ports mean 2.19-in. intake valves.
6. Before 1970, 1.72-in. exhaust valves were used in all cast-iron heads.
7. For 1971 and after, 2.19-in. intakes mean 1.88-in. exhaust valves.

Aluminum heads don't follow so many rules. Here are the few:
1. All 1967—'70 closed-chamber heads use 2.19-in. intakes and 1.84-in. exhausts.
2. All open-chamber aluminum heads use 2.19-in. intakes and 1.88-in. exhausts.
3. All aluminum heads are large-port heads.

Unfortunately, it is not possible to predict open- and closed- chamber heads with any such rules. What you can remember though is; closed chambers were used exclusively until 1968. They gradually died out as the cleaner-running, open-chamber heads took over.

ing guide inserts—as described in the cylinder-head chapter. If you want replacement guides, they are available from TRW or Manley.

Sparkplugs—A quick way to tell an "old" head from a "new" one is by the sparkplug type used. Iron heads, up to 1970, use "conventional" 3/4-in.-reach, 13/16-in.-hex, gasketed plugs. Through 1970 and afterwards, 5/8-in. hex, taper-seat plugs were used exclusively. These allowed larger water jackets for improved cooling on the hotter-running emissions engines. All aluminum heads use 13/16-in. gasketed plugs.

43

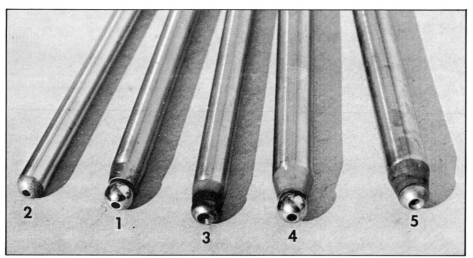

(1) 1965—'66 3/8-in. pushrod with welded-on ends; (2) The 5/16-in.-diameter pushrod was used from '67 to '71 in hydraulic-lifter engines; (3) Replacement 3/8-in. pushrod was meant for use with mechanical lifters; (4) 1st-design L-88 pushrod features ball tips welded to 7/16-in.-diameter tube. (5) 2nd-design L-88 uses 7/16-in. pushrod with pressed-in-ball tips. Always replace worn pushrods with *new* ones. Order by year and HP rating. Don't worry about the tips—concern yourself diameter and length. Photo by Bill Fisher.

A final note about sparkplugs: Gasketed plugs for cast-iron cylinder heads are partially threaded; those for aluminum heads are fully threaded. Fully threaded plugs give additional thread engagement as insurance against stripping the aluminum threads.

A look at the sparkplug seats will give you a rough idea of when a cast-iron head was made. Flat plug-sealing surfaces mean pre-'70; tapered seats indicate 1970 or later head.

Water-Temperature Sender—Another head-identification guide is the location of the water-temperature sending unit. Pre-'68 big-blocks have the sending unit mounted in the intake manifold, next to the thermostat housing. On '68 and later engines, the sending unit is in the left cylinder head, between the center exhaust ports. If you have an early engine and decide to run the later heads, you'll need Chevy's threaded plug 444667 or 444746 for both heads.

PUSHRODS

Three different-diameter pushrods have been installed with three different-style tips. See photo on this page. Matching the three pushrod diameters are three different guide plates. The slots in the guide plates match the diameter of the pushrods used. And intake and exhaust pushrods are different lengths: 8-7/32 in. for the intakes; 9-3/16 in. for the exhausts. The "taller" truck pushrods are 0.4 in. longer.

All '65 and '66 big-blocks use 3/8-in.-diameter pushrods with welded-on steel-ball tips. This type of pushrod is also standard on all mechanical-lifter engines through 1970. In 1970, a replacement pushrod was introduced for solid-lifter engines. While it also measures 3/8 in. it has different tips. Instead of being welded on, the steel balls are pushed halfway *into* the ends of the tube.

Hydraulic-lifter big-blocks produced from 1967 to 1969 use smaller, 5/16-in.-diameter pushrods. These pushrods have simple, closed ends similar to those used in small-block Chevys.

High-performance big-blocks from 1967 on have 7/16-in.-diameter pushrods. First design L-88 engines—1967 only—have the large pushrod with welded-on steel-ball tips. With the advent of the second-design L-88 in 1968, the pushrod ends were reworked to accept the pressed-in-ball tips. If you have a 7/16-in.-pushrod engine, the only service pushrods from Chevy have the later, superior, pressed-in tips.

From 1970 to mid-'71, 402s reverted to the early 396 3/8-in. replacement pushrod with pressed-in tips. Halfway through 1971, the 402s got the smaller 5/16-in. pushrod. All 402s built after this date use the 5/16-in. pushrod.

The 454 went through similar pushrod-usage changes. In 1970 the low-performance LS-4s got the 5/16-in. pushrod. Everything up to the LS-6 uses the 3/8-in. pressed-in-tip pushrod. The LS-7 sports 7/16-in. pushrods of the second L-88 design—pressed-in-ball tips. In 1971 and '72 all 454s got the 3/8-in. pushrod, except for the LS-7. Then, from '73 on, all 454s returned to square-one; they use 5/16-in. pushrods.

If you need to buy pushrods, I recommend that you get a full set. Make sure the parts man delivers the right ones by lining the pushrods up on the counter and counting them out: eight long and eight short. And don't forget: *If pushrod diameter is changed, the the guide plates must also be changed.*

MANIFOLDS

Intake—Other than the wider 366- and 427-truck manifolds, important design features of big Chevy intake manifolds are limited to two areas: runner size and material. To mate with either large- or small-port heads, there are large- and small-runner manifolds. Some high-performance engines had aluminum intake manifolds; all other manifolds are cast-iron.

Although you may have never seen anything other than a single 4-barrel carburetor on a big-block, there are two other styles. In 1969, there was a 265-HP 2-barrel-carb big-block for passenger cars; '67—'68 Corvettes had optional three 2-barrel Holleys.

Chances are your big-block has a 4-barrel Rochester Quadrajet carburetor on a low-rise, cast-iron intake manifold mated to small-port heads. This is a good combination, but you may wish to make some changes. The important considerations are: bolt patterns, port configuration and plenum division.

Bolt Patterns—Along with the Quadrajet, Chevy also offered 4-barrel Holley carburetors. These do not interchange because of different bolt patterns. Your Q-Jet manifold can be equipped with a Holley Model 4165 or 4175 spread-bore carburetor or a Carter Thermo-Quad 9800. These fit onto any manifold designed for the Q-Jet. If you have a truck or tow vehicle that spends a lot of time at part-throttle, using a Q-Jet or other spread-bore replacement is a good way to maintain good driveability, with acceptable performance. Avoid installing any carburetor with an adapter.

When interchanging manifolds, remember that the new one must match the cylinder-head ports. A small-port manifold will not fit large-port heads *correctly,* and vice versa. This is especially true with a large-port manifold on a small-port head. The large-to-small transition at the head-to-manifold interfaces will be detrimental to power output; more so than small to large.

Except for 366 and 427 trucks, any big-block manifold will *bolt* to any big-block engine. Most manifolds you'll find at the junkyards are likely to be for Quadrajets and have small-runners; look for the oval ports at the manifold faces in a cast-iron part.

A 4-barrel aluminum manifold *usually* means it has large ports, regardless of runner shape. I say ''usually'' because mid-'70 big-blocks in Vettes with Q-Jets have *small-port* aluminum manifolds. Three 2-barrel aluminum intakes come with both large and small ports. The small ports are oval; large ports are rectangular.

Something to keep in mind when manifold shopping is whether the manifold you are looking for will accept all the plumbing required for your engine. Early intake manifolds were drilled for only *one* vacuum fitting. A tee at the single hole routes vacuum to the power-brake booster and heater/air-conditioning vacuum motors.

Later big-blocks are a different story. Exhaust-gas recirculation (EGR) fittings, accessory-mounting holes, different choke systems and relocated water-temperature senders mean different holes in the intake manifold. There's no way I can list the various big-block manifolds. But if you're looking for a replacement intake manifold, make sure any prospects will accommodate the accessories and fittings needed for your engine. Extra fittings are no problem; just remove 'em and put plugs in the holes. On the other hand, if the manifold doesn't have the necessary tapped hole/s, drilling them may not do the job as *bosses*—additional material—are cast-in where fittings are to be installed. This is to give additional strength and sufficient thread length.

Truck Manifolds—Adapters are available to mate *large-port* passenger-car intake manifolds to the wider 366- and 427-truck blocks. Weiand's

Aluminum intake manifold at top is large-port design used with Holley carb on LS-6 and LS-7 454s. Manifold at bottom is the more common small-port 454 for use with a spread-bore Holley. Note difference in carb-mounting shapes. Both manifolds are fitted with a sheet-metal shield underneath to keep crankcase oil from being overheated by the exhaust-crossover passage. Photo by Bill Fisher.

Only difference between these two aluminum, large-port, Holley-carb manifolds is plenum-divider height. Manifold at top with its cut-down plenum is not suitable for the street, so don't install one thinking it is the hot-tip. Fully divided-plenum manifold is what's needed for top street performance. Photo by Bill Fisher.

Tri-power was a 427-Corvette option from '67 to '69. This is a large-port manifold, but a small-port version was also available. The center carburetor is the equivalent of the primary side of a four-barrel; end carburetors are vacuum-operated. Only the center carb has a choke. These Corvette-induction setups are valuable in today's collector market. Photo by Bill Fisher.

A real find at the swap meet, a three 2-barrel manifold. All that's needed now are the carburetors, linkage and air cleaner.

spacer-plate set 8204 sells for about $60. It will be a lot easier and less expensive to use the original manifold, if possible. As for installing a 366- or 427-truck manifold on a passenger-car big-block, there's no way.

Plenum Dividers—Inside all big-block manifolds, under the carburetor is a vertical wall, or *plenum divider,* separating the two runner *planes.* This divider greatly assists low-rpm torque. When partially cut down, high-rpm power output increases. But this is *at the expense of low-end torque.* An engine with a manifold so modified is sluggish at low rpm, but runs well at high rpm. Great for super speedways, but awful around town! An *open-plenum* manifold is not suitable for any kind of street driving. Even hot-rodded street engines rarely see the high-rpm use that would justify such a modification.

Tri-Power—The *tri-power,* or three 2-barrel, Corvette option is not only impressive looking, it runs well. Forget about tracking down such a setup. Not only are tri-power setups very expensive, they are rare. And although a tri-power-equipped engine makes "good" horsepower, a single 4-barrel can do better.

So if you acquire a tri-power setup, sell it to a Corvette big-block restoration enthusiast. You'll get top-dollar and the parts will end-up where they belong—on a Corvette. As for the money, you'll have more than enough to buy a better carburetion system.

Exhaust Manifolds—There is little choice when it comes to swapping exhaust manifolds. The problem is the lack of engine-compartment room. Undoutedly, you've come to realize that the big-block Chevrolet is a *big* engine! There isn't much room for anything more under the hood—including larger exhaust manifolds. Here is one item Chevy compromised to make the engine fit the various engine compartments.

The choice boils down to two basic big-block exhaust manifolds: passenger car and Corvette. Passenger cars and trucks got a more-restrictive *log-type* cast-iron manifolds. Corvettes were blessed with less-restrictive manifolds that add up to more horsepower and better gas mileage!

In case you are thinking of shoehorning the Corvette manifolds onto your Impala or El Camino—they don't fit. Stick with the manifolds that came with your engine. If your car has a single exhaust system, you can install a dual system for reduced back pressure and a little relief at the gas pump.

Exhaust Headers—Aftermarket, tube-type exhaust manifolds will improve mileage, produce a little more horsepower, cost a fair amount of money, make a lot more noise and need replacing after a relatively short time—you make the decision. If you opt to stay with the stock manifolds, you may miss out on *possible* more miles-per-gallon and a little extra power.

However, stock cast-iron exhaust manifolds should last forever—unless you ford rivers. Cold water splashed

Late-model 454 exhaust manifold is tapped and fitted with air-injection tubes. Shield protects spark-plug leads from exhaust-manifold heat.

Corvette exhaust manifolds are less restrictive than their passenger-car and truck counterparts. Except for big trucks, Corvette manifolds fit only Corvettes, so there are few swaps to be made here. Photo courtesy of Chevrolet.

on hot cast-iron headers will crack them. You will also enjoy a quieter-operating power plant, all things being equal. On the other hand, you will have to live with more noise from tube-type headers. Drive another big-block so equipped before laying out the green for your own set.

Headers or not, be sure you get manifolds with air-injection fittings if the engine has a smog-pump—*A.I.R.* in Chevy parlance. The men in blue will get you sooner or later if you bypass smog controls—and the fine isn't light in some states. A smog check may be mandatory when changing titles or getting next-year's license. Do your part to keep our air clean. Keep the A.I.R. system operational with tapped manifolds.

Other exhaust-manifold considerations include manifold shrouds, or *stoves,* for carburetor heat and provisions for exhaust-heated chokes. Both of these items are in the "if-you-need-'em, get-'em" category. The stove is that sheet-metal box riveted to the right-side exhaust manifold on later cars and trucks. It collects hot air from around the manifold and sends it up a flexible tube to the air cleaner. Because the riveted connection is hard to duplicate, make sure the manifold you buy has the sheet metal attached, if required for your engine.

Early exhaust-heated chokes were found on L-78 396 and 402 engines. They consist of a length of metal tubing to route exhaust gasses from the exhaust manifold to the choke. If you have an L-78, find an exhaust manifold tapped for the metal tube, or change the choke to the later style that gets its exhaust heat from the crossover passage in the intake manifold.

Another alternative is to use an aftermarket carburetor with an electric choke.

BIG-BLOCK PASSENGER-CAR RPO DESCRIPTIONS

Grouped by displacement and Regular Production Option (RPO) Number, and in order of increasing performance.

396 CID (4.094 x 3.76 Bore & Stroke)

L-66
1969 only
265 HP 2-Bbl carburetor, low-lift hydraulic camshaft, small-port cast-iron heads, 9:1 CR, 2-bolt mains, nodular-iron crankshaft.

L-35
1965-69
325 HP Holley or Q-Jet 4-Bbl carburetor (Q-Jet only after 1967), low-lift hydraulic cam, small-port cast-iron heads, 10.25:1 CR, 2-bolt mains, crankshafts: forged 1965-67; nodular iron 1968-69.

L-34
1966-69
360/350 HP Holley 4-Bbl carburetor until 1968, then Q-Jet, 1968 and 69, high-lift hydraulic cam, small-port cast-iron heads, 10.25:1 CR, 2-bolt mains (some 1966 Chevelles had 4-bolt mains), forged crankshaft.

L-78
1965-69
375 HP (rated 425 HP in 1965 Corvette only) Holley 800 CFM 4-Bbl carburetor, mechanical-lifter cam, large-port cast-iron heads, 11:1 CR, 4-bolt mains, forged Tuftrided crankshaft, forged pistons. Sold in Chevelle, Chevy II and Camaro.

L-89
1968-69
375 HP Limited edition of L-78 engine produced with non-open-chambered aluminum cylinder heads in Chevy II and Camaro.

402 CID (4.126 x 3.76 Bore & Stroke)

LS-3
1970-71
330 HP (300 HP 1971) Q-Jet 4-Bbl carburetor, low-lift hydraulic cam, small-port cast-iron heads, 10.25:1 CR (1970) and 8.5:1 CR (1971), 2-bolt mains, nodular-iron crankshaft.

LS-3
1972
240 HP w/dual exhaust, 210 HP with single exhaust, other specs same as 1970-71 LS-3.

L-34
1970 only
350 HP Q-Jet 4-Bbl carburetor, high-lift hydraulic cam, small-port cast-iron heads, 10.25:1 CR, 2-bolt mains, forged crank.

L-78
1970 only
375 HP Identical with L-78 396 except for bore size.

427 CID (4.251 x 3.76 Bore & Stroke)

LS-1
1969 only
335 HP Q-Jet 4-Bbl carburetor, hydraulic camshaft, small-port cast-iron heads, 10.25:1 CR, 2-bolt mains, nodular-iron crank.

L-36
1966-69
390/385 HP Holley or Q-Jet 4-Bbl carburetor, high-lift hydraulic camshaft, small-port cast-iron heads, 10.25:1 CR, 2-bolt mains, forged crank.

L-68
1967-69
Corvette
400 HP Identical with L-36 except for 3 Holley 2-Bbl carburetors on small-port aluminum intake manifold.

L-72
1966-68
425 HP Holley 800 CFM 4-Bbl carburetor, mechanical camshaft, big-port cast-iron heads, 11:1 CR, 4-bolt mains, forged Tuftrided crank, forged pistons, aluminum high-rise intake manifold.

L-71
1967-69
Corvette
435/425 HP Identical with L-72 except for 3 Holley 2-Bbl carburetors on aluminum manifold.

L-89
1968
Corvette
Similar to Corvette L-71 except for aluminum cylinder heads (non-open-chamber-type).

L-88
1967-69
Corvette
430 HP Holley 850 CFM 4-Bbl carburetor, high-lift mechanical cam, aluminum cylinder heads (open chamber in 1969), 12.5:1 CR, 4-bolt mains, forged Tuftrided crank, forged pistons, floating wrist pins, 1/16-in. compression rings, 7/16-in. pushrods, open-plenum intake manifolds.

ZL-1
1969-70
Corvette
430 HP Identical features to the L-88 engine except all-alluminum cylinder block.

454 CID (4.251 x 4.00 Bore & Stroke)

LS-4
1970
345 HP Q-Jet 4-Bbl carburetor, low-lift hydraulic camshaft, cast-iron small-port cylinder heads, 10.25:1 CR, 4-bolt mains, forged crankshaft.

LS-4
1973-74
215/235/245/270 HP Q-Jet 4-Bbl carburetor, low-lift hydraulic camshaft, cast-iron small-port cylinder heads, 8.95:1 CR, nodular-iron crankshaft.

LS-4
1975-76
215/225 HP All specs same as 1973-74 LS-4 except for CR lowered to 8.85:1.

LS-5
1970-72
360/390 HP Q-Jet 4-Bbl carburetor, high-lift hydraulic camshaft, small-port cast-iron heads, 10.25:1 CR, 4-bolt mains, forged crankshaft.

LS-5
1971
365 HP Specs as per 1970 LS-5, except for small-port open-chamber cast-iron heads with 8.5:1 CR.

LS-5
1972
270 HP with dual exhaust, 230 HP with single exhaust, other specs same as 1971 LS-5.

LS-6
1970
460/450 HP Holley 800 CFM 4-Bbl carburetor, mechanical cam, cast-iron large-port cylinder heads, 11:1 CR, 4-bolt mains, forged Tuftrided crankshaft, forged pistons with pressed-in pins.

LS-6
1971
Corvette
425 HP Holley 800 CFM 4-Bbl carburetor, mechanical cam, aluminum large-port open-chamber cylinder heads, 9.0:1 CR, all other specs as 1970 LS-6. Cast-iron large-port open-chamber heads available as service parts.

LS-7
1970
Corvette
465 HP Never released for production by Chevrolet. Consisted of Holley 850 CFM 4-Bbl carburetor, mechanical cam, large-port cast-iron heads (open-chamber), 12.5:1 CR, 4-bolt mains, forged Tuftrided crankshaft, forged pistons, pressed wrist pins, 1/16-in. compression rings, 3/8-in. pushrods, open-plenum intake manifold. May be assembled from available Chevy service parts.

1972 power ratings were reduced because GM changed to an SAE 85F correction factor rather than the 60F factor previously used for the gross-power-rating method.

4 Teardown

Tearing down an engine goes quickly, but don't go *too* quickly. Take time to look for signs of wear as you go. Too many mechanics scatter parts in record time, losing valuable clues that could help with parts inspection and reconditioning. Don't fall into this trap. Study internal-wear patterns as you remove each part. You'll be able to spot the problem areas in your engine and estimate the work and parts required to fix them. Teardown is an opportunity to observe the effects of service, or lack of it—and to discover the cause of that clacking noise or trail of blue smoke.

Follow the same basic procedure you used when pulling the engine—separate hardware according to function and use labeled containers for storage. Pay special attention to the location of internal parts and their positions relative to each other. This is your last chance to see them in their factory-installed order—if this is the engine's first rebuild. Sketches or photographs can be very helpful, but make them *before* you remove the parts.

Big-block Chevrolet supplied the power demands of the late '60s. Early 427 sans cumbersome accessories and emissions-control devices. Photo courtesy of Chevrolet.

EXTERNALS

While the engine is hanging from the hook, remove the flexplate. If there's a clutch and flywheel, *index* the pressure plate to the flywheel and the flywheel to the crankshaft *before removing them.* Use a hammer and center punch to put three dimples in each piece, adjacent to one another.

Index marks will let you reassemble the clutch and flywheel in their same relative positions so engine balance will be maintained. Although theoretically unnecessary, as these parts are balanced separately—with the exception of the 454 counterbalanced flexplate or flywheel—indexing them will eliminate any doubts.

Engines mated to automatic transmissions do not require a flywheel. In place of the flywheel is a *flexplate.* The flexplate drives the torque converter and supports the starter ring gear. The torque converter is heavy enough to fulfill the flywheel function.

Flexplates are removed the same as a flywheel: Remove the six mounting bolts and pull the plate free. As shown in the photo, page 50, a pry bar or large screwdriver slipped through a lightening hole will keep the crank from turning. When handling the flexplate, beware of the sharp edge around the flexplate center hole. I got a pretty bad cut once when I reached through this hole to pick up a flexplate. Be careful. If your flexplate has any sharp edges, remove them with a file before they "bite" you.

Clutch & Flywheel—After matchmarking the pressure plate, loosen each of its bolts half a turn. Repeat this until the pressure plate is loose. This will keep the pressure-plate cover from being warped. As you remove the last two bolts, be ready to grab the pressure plate and disc before they fall.

Keep your dirty hands off the clutch disc. Grease on the facing material will make the clutch grab and chatter. One greasy fingerprint can do it, so be careful.

Now for the flywheel bolts. If you own an impact wrench, use it. If not, use a breaker bar and socket. You'll find that the crankshaft turns as you attempt to break the bolts loose. To solve this problem, use another socket and breaker bar or ratchet at the damper bolt. Or put your brawniest screwdriver to work. You'll need a friend to hold the screwdriver. Place the screwdriver blade between two ring-gear teeth and rest its shank against a bellhousing dowel or a bellhousing bolt threaded back into the block. With the crank held from turning, break loose all flywheel bolts. Remove them and lift off the flywheel.

Water Pump—At the front of the engine, remove the thermostat-bypass hose from between the water pump and the intake manifold. Remove the four bolts holding the pump to the block. A light jerk on the large water inlet should break a sticky pump free. Beware of trapped coolant cascading from the block. If the pump is really stuck, a few light raps with a mallet on the pump-housing nose—*not the shaft*—should break it loose.

Drain Liquids—To avoid a big mess, you'll need a drain pan for catching coolant or crankcase oil. This could preserve your good relations with family, friends and neighbors.

Finish draining trapped coolant by removing the two hex-head drain plugs low down on each side of the block. Try a box-end wrench first. If the plugs won't budge, try using an adjustable wrench. It should have enough length for the needed leverage. Be careful that you don't round off the hex. If no coolant comes out after removing a plug, poke a wire through the hole to clear the sediment.

Remove the oil-pan drain plug to drain the crankcase oil. Remove the oil filter while you have something to drain the oil in.

Core Plugs—There are three core-plugs in each side of the block, plus two in the rear face. These plugs rust from the inside. Even though they look OK, replace them! They are very difficult to replace once the engine is back in the chassis. To remove a core plug, knock it in with a hammer and a large, blunt punch. Grab the edge of the plug and lever it out with Channel-lock pliers.

Thermostat Housing—Back on top of the engine, remove the two bolts that retain the thermostat housing. If it sticks, run the wrench handle into the housing neck and pry it off—not too hard or you'll break the neck. With the housing off, lift the thermostat from its hole.

Ignition Coil—You'll find the coil at the left rear of the intake manifold. To remove the coil, disconnect the small wire connecting the distributor to the negative terminal of the coil. Then remove the bracket bolts, including an intake-manifold bolt, and remove the coil. If your engine has High Energy Ignition (HEI), the coil is integral with the distributor cap. In this case, remove the distributor cap and plug wires.

Engine Stand—It's time to get the engine where you can finish tearing it down. Although a strong bench is suitable, I think the best way to support a big-block Chevy for teardown and assembly is with an engine stand. Although it is a large investment, I find an engine stand pays off in time and convenience. You can buy a new or used stand, or borrow or rent one. A word of caution: Make sure the stand can handle the load—it will have to support up to 700 lb. Even the brawniest stand will groan under this load, so don't borrow one from a Volkswagen garage!

Then there is always a work bench: Although I like an engine stand, you might prefer working on a bench. I

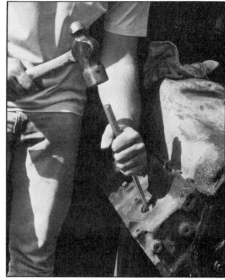

Remove core plugs even though they look good. Unless they are rusted through, you can't tell their condition from the outside; they rust from the inside out. This will also drain the block if you didn't remove the drains, so watch out. Knock each plug in sideways, then pry it from block using pliers.

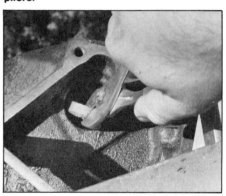

Remove some components before you put engine on a bench or engine stand. Start with the flexplate or flywheel. A wrench on the crank-damper bolt or a screwdriver on a bellhousing dowel and flywheel/flexplate tooth will also keep crank from turning. Next, go to the front of engine and remove water pump.

Drain coolant trapped in block. A block drain is provided below center core plug on each side of block. With a pan underneath, remove plugs. You may have to poke a screwdriver or whatever through the holes to clear the blockage.

Good and bad practices are illustrated here. Segregating fasteners in small containers is good; mixing rocker arms and their ball pivots will result in high wear later. Photo on page 54 shows how to keep the rockers and balls paired.

Remote choke with rod-type linkage, left. Remove clip to disconnect rod at carburetor. Integral choke, right, senses exhaust gasses piped to choke housing. Disconnect tube at choke.

Why wrestle around a heavy engine such as the big Chevy on the floor or a bench while rebuilding it? Get an engine stand such as this Lakewood 33010 that has the beef needed to support a heavy engine.

Wrong! A flare-nut wrench should be used to prevent rounding the flare nut. Otherwise, everything is as it should be. Squeezing the wrenches together should break the fuel-line-to-carburetor connection with little effort.

know one professional mechanic who won't use anything else! If you choose the bench route, make sure you use a sturdy one.

Carburetor—Use a 1-in. open-end wrench and 5/8-in. *flare-nut wrench* to disconnect the fuel line at the carburetor.

Automatic chokes are controlled by a *bimetal spring*. When the bimetal spring is heated by exhaust gas the choke opens; when cooled, the spring closes the choke.

Two basic types of automatic chokes are used: *remote* and *integral*. If yours has a remote choke, the bimetal spring is in a cavity above the intake-manifold exhaust-crossover passage. It operates the choke through a short rod. Unclip this rod at the choke-plate shaft.

An integral choke is easily identified by the cylindrical housing at the end of the choke-plate shaft. The bimetal is enclosed in the housing and operates the choke *directly*—as opposed to remotely. Exhaust-gas heat is routed through a steel tube to the choke housing from one of two sources: the exhaust manifold or the intake manifold. The tube from the exhaust manifold is long: the one from the intake is a short loop under the choke housing.

Disconnect the tube at the choke. Label and disconnect the vacuum-advance line, plus whatever other vacuum lines the carburetor may have connected to it. Remove the four nuts—or bolts—that secure the carburetor to the manifold. Lift the carburetor off the intake manifold.

If your engine is equipped with a Rochester Q-jet, it's mounted with four bolts; two long and two short. At the front, the long bolts pass through the top of the carburetor and into the intake manifold. The short mounting bolts extend through the mounting flange at the rear of the carburetor base. Holley carburetors use four equal-length studs with nuts at the base of the carburetor.

Take care not to set the carburetor down on the throttle plates, you'll bend them.

Distributor—Remove the sparkplug leads from the plugs. Take care not to yank the leads off. Give them a twist when pulling off the boots. Most engines use slip-on clips, so all you need do is pull them off the rocker covers. Others use metal stands anchored under a rocker-cover bolt. Remove the bolts to remove the clips. Replace the rocker-cover bolts so you won't lose them. Leave the clips on the leads; they'll help relocate the wires during engine reassembly.

Handle the sparkplug leads carefully. Except for solid-core leads, ignition "wires" are really nothing more than carbon-impregnated threads. Rough handling will break the threads and increase the lead's resistance until it can't fire the plug.

Remove the wires and distributor cap as a unit. Point-type distributors have two J-hooks securing the cap, HEI units have four. To release their grip, press down with a screwdriver and turn them 1/4-turn to the left, or counterclockwise. Pull up on the cap to remove it and the wires.

Studs and nuts or bolts secure carburetor to manifold. Two-barrel Rochester uses a stud/nut combination at each corner. Holley four-barrels are the same way. Rochester four-barrels have a bolt at each front corner that passes through carb and threads into manifold; shorter bolts are used at rear.

Try to keep carburetor and linkage together. This will keep the small parts from getting lost—and it saves time.

After removing distributor cap and spark-plug leads, remove distributor-clamp bolt and clamp. Lift distributor out of engine.

To provide radio shielding not possible with the Corvette's fiberglass body, a metal box is installed over distributor. Additionally, a ground wire for each shielded sparkplug lead is installed under each rocker-cover bolt in the bottom row (arrows). Remove bolts and disconnect ground wires.

After you remove engine mounts, inspect them. If they are cracked or separated from the steel like this, replace them.

Remove the distributor hold-down clamp; it's at the base of the distributor. Some big-blocks use a bent-wire clamp—others a small, stamped-steel clamp. Remove the bolt and hold-down. The distributor can now be pulled up and out of the intake manifold.

Exhaust Manifolds & Things—Turn the engine over on its right side and remove the left exhaust manifold. Eight 9/16-in. bolts will have to come out first. If your engine has heat shields around the sparkplugs, remove each by undoing the bolt directly above the plug.

On '68 and later models, the water-temperature sending unit was installed in the left head; remove it now. These sending units are fragile, so be careful. If you crack the center insulator, the unit will have to be replaced.

With the same wrench you used on the exhaust-manifold bolts, remove the three engine-mount bolts and the mount. Inspect the metal-pad-to-rubber bond, as well as the rubber, for cracks and separations. If the rubber is peeling away from the metal, or is soft from oil contamination, replace the mount.

You'll find the oil-pressure sender on the left side; on some engines it is

Make sure all 16 intake-manifold bolts are out before you try to break loose manifold. Some manifold bolts do double-duty; this one retains the A/C-compressor bracket. Wedge between block and manifold to break it loose. Pry only along this flat area at front or rear of engine.

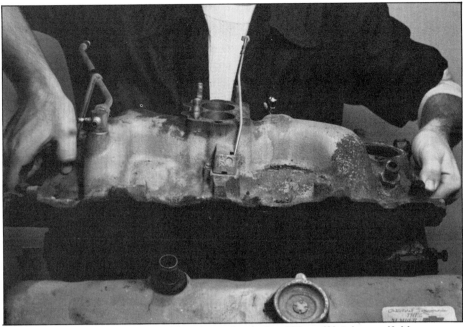

Unless it's aluminum, be prepared for a heavy load when lifting off intake manifold.

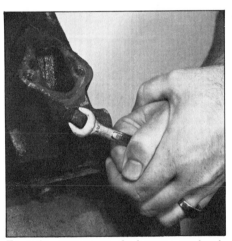

To gain access to fuel-pump pushrod, remove this plug. Fish pushrod out of hole, below.

Unless you have objections, remove fuel pump with pump-to-carburetor fuel line. Otherwise, remove pump with a flare-nut wrench. Then remove the two fuel-pump bolts and pump.

Remove the six bolts each and pop rocker covers off. A sharp tap under cover flange should do it.

near the bottom of the block. Remove it before it gets damaged.

Roll the engine over on its left side and remove the right exhaust manifold, heat shields, engine mount and dipstick tube. After it's unbolted from the exhaust manifold, pull the dipstick tube out of the short pipe at the oil pan. On the bottom of the tube is an O-ring—don't lose it. Inspect the right engine mount.

Intake Manifold—Go back on top and remove the intake manifold. Later engines became increasingly complex. They are covered with what seems to be miles of vacuum tubes and hoses. Remove these *only* after you've labeled them. Make doubly sure you'll be able to reinstall them properly by taking some pictures. *This is important.*

There's a row of 9/16-in.-hex bolts running along each side of the manifold; remove them. The manifold will have to be pried off. *First, make doubly sure you have all the bolts out!* Use a medium-size screwdriver to pry

between the manifold and the block.

With a little leverage you should be able to pry up one end. Once it's broken loose, grip the manifold with your fingers, lift it off and set it aside. The lifters and pushrods will now be exposed.

Cylinder Heads—Now for the heads.

Water-temperature sender should be removed before head. If you leave it in, chances are it will get broken before the rebuild is completed.

Removing rocker-arm can be a chore. Prevailing-torque nuts, they should be tight all the way off. If you find any that aren't, replace them. Otherwise, valve clearance won't be maintained.

Remove and store each rocker arm and ball as a unit.

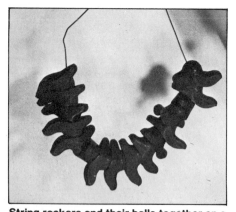

String rockers and their balls together on a wire as I've done with this coathanger to keep them paired and in order. You can clean and inspect them without removing them from the wire; they don't have to be removed until final assembly.

Pull pushrods out and store them in order. How you store them needn't be fancy. Punch eight holes in two pieces of cardboard and indicate FRONT or REAR of head and RIGHT or LEFT side of engine. Simple!

Working one side at a time, unbolt each rocker cover. Remove it with a light tap with the heel of your hand or a soft mallet. If it sticks, pry lightly under the gasket flange at a corner with a screwdriver.

Then, with a long ratchet for extra leverage, start the laborious task of removing the *prevailing-torque* rocker-arm nuts. The top portion of each nut is triangular in shape so it clamps onto the rocker-arm stud and maintains valve adjustment. These 11/16-in. demons nestle down in the center of the rocker arms and will be tight all of the way off.

Next, remove each rocker arm with its fulcrum; *keep them together*. String them on a wire so they don't get lost or mixed up. Store the rocker arms, their fulcrums and nuts in a rocker cover. Because a rocker-arm and its fulcrum wear-in together, they must

be kept together. Otherwise they'll wear rapidly or, possibly gall.

Pushrods are next: Grasp each one and wiggle it to free it from its lifter. Pull it out and insert it into a cardboard holder as shown. As with the rockers and their fulcrums, you've got to keep the pushrods, rocker arms and fulcrums *paired* and in order, or they'll wear rapidly.

Before you proceed with removing the heads, try removing the lifters, or tappets. *Keep them in order.* If you can pull them out of the top of their bores easily, they will fall out when you roll the engine over. Their order will be lost; a financial setback if the cam and lifters are reusable. Read about the importance of lifter order on page 80 in the next chapter.

You're now ready to remove the heads. Make sure the engine-stand neck is tight or, if you're working on a

bench, support the engine so it can't roll over. If it's teetering now, it'll roll over for sure when you lift off a head and engine weight shifts. So make sure the engine is stable before you proceed with removing the heads.

There are three rows of head bolts: The top row is—or was—under the valve cover, in the area of the valve springs and pushrod guide plates. The bolts in the middle row straddle the exhaust ports. The bottom bolts are along the bottom edge of the head, under the exhaust ports.

You'll need a breaker bar to break the head bolts loose. Once they're broken loose, run 'em out with a ratchet. Before starting to pry the heads loose from the block, make sure *all* head bolts have been removed. Count them; if you don't have 16 for each head, you've missed one or more. If you pry hard enough

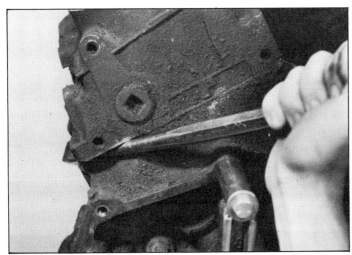

After you've removed *all* 16 bolts from each head, thread a head bolt or two back in to keep a head from falling onto the floor or your foot. Break the heads loose from the block. Pry at the lower, rear corner between the head and block. If the head is stubborn double-check to make sure all head bolts are out. Above all, *don't wedge anything between the block and head* or you'll damage their mating surfaces.

Before you lift off a head, make sure engine can't roll over as its center of gravity shifts. Also, make sure it isn't too high or low for your height. Remove "safety" bolts. Get a good grip and lift head straight up off of its dowels.

you could break the head—strong as it is—so don't feel foolish about double-checking. Each head is now "sitting" on the block, held tight by its gasket.

You'll need a pry bar to break the heads loose from the block. But first, thread two head bolts back in each head; two turns will do. Doing this will keep a head from falling off onto the floor or your foot when it breaks loose. If a head falls on your foot, you could be hospitalized!

Pry between the head and the small ledge at the lower, rear corner of the block as shown. Once you've got a head unstuck, remove the bolts, lift the head off and set it aside. Be ready to handle a heavy load. A cast-iron big-block head weighs about 70 lb. Big-block Chevy heads are so heavy that some machine shops charge extra just for handling them! The point is, be careful when handling the heads—and any other big-block components.

Repeat the process with the other cylinder head. Once it's off, you're finished with the cylinder heads for now. Cylinder-head teardown, inspection and assembly is covered in Chapter 6.

Harmonic Balancer—Getting into the *bottom end* begins with disassembling the front of the engine. Start with the crankshaft damper. Remove the 3/4-in.-hex bolt and *two washers* that retain it to the crank snout; a lock washer and a heavy flat washer are used behind the bolt. The flat washer

Unless you are using an impact wrench, you'll have to keep crank from rotating while attempting to loosen damper bolt. I've threaded two flywheel/flexplate bolts back in crank flange and bridged them with a pry bar to accomplish this.

is quite large. You might mistake it for the crank snout if it sticks to the damper. Remove it.

To hold the crank when breaking the damper bolt loose, thread two flywheel or flexplate bolts back into the flange at the rear of the crank. Bridge these bolts with a prybar or large screwdriver to keep the crank from turning.

You'll need a puller—borrowed, bought or rented. **Do not use a pulley puller**—its jaws hook around the back of the part being removed. This type of puller will destroy the harmonic balancer. The outer ring that is bonded to the center with rubber will tear loose. Rather, use a puller that pulls against bolts threaded into the part being removed.

Remove damper bolt, lockwasher and heavy flat washer. Thread bolt back in, less flat washer, so center puller bolt will have something to bear against. Damper will clear the bolt head as it comes off.

Run the damper bolt back into the crank center so the puller screw will have something to push against. Now, attach the puller to the three threaded holes in the damper. Take care to align the puller plate to the damper. You need a straight, forward pull.

Run the puller bolt in against the damper-bolt head to pull the damper off. When the damper loosens, it should pull off by hand. Remove the puller from the damper and the damper bolt from the crankshaft.

If the damper is stubborn and you are sure the puller is applying a strong, *even* pull, try a *little heat* from an acetylene or propane torch. With the puller tight, apply the heat *only to the damper center,* or hub. Don't apply too much heat. With tension on the

Use this type of puller to remove damper, not a pulley puller that hooks around piece it removes. Run center puller bolt in against damper-bolt head and pull off damper.

As with most high-mileage timing chains, this one is shot. Note how slack the far side is.

Turn crank or cam to tighten one side of chain. If slack side deflects more than 1/2 in. at midpoint, it's junk. Replace it.

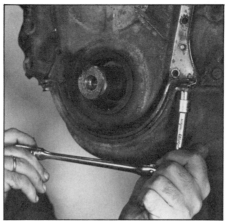

Remove front-cover bolts. Make sure you remove the two or three bolts that thread into front cover from underneath. They secure oil pan to the front cover. Pry against the back of the front-cover flange to break it loose.

damper, it should pop loose when heated.

Harmonic dampers absorb vibration through the rubber that bonds the outer ring to the damper hub. Heat the hub, but not so much that you damage the rubber. If the rubber ring is weakened, it could cause the damper to fail after the engine is back in operation.

Front Cover—Use a ratchet and a 7/16-in. socket and undo the front-cover, or timing-cover, bolts. There are 10 around the upper perimeter of the cover; two or three more hide under the oil-pan lip. If three bottom bolts are used, one is in the bottom center and the other two are at each outer corner of the oil pan. Count the bolts; if you don't have all 13, don't try to remove the cover.

The cover is stuck on by gaskets, just like everything else. Unlike the heads, the cover is sheet metal and will bend easily. Start prying carefully at the top of the cover and work around each side to the bottom. Once the cover starts to peel away from the gasket, it should come off easily.

With the front cover off you'll be able to peer into the crankcase. No big deal unless you are anxious to confirm whether or not you have a four-bolt block. Enough of the number-1 main cap will be visible so you'll be able to tell whether it has two or four bolts. It's dark in there, so you may need a flashlight.

Inspect Timing Chain—Because excess timing-chain and sprocket wear is obvious, there are few cases where detailed inspection of these

Remove the three cam-sprocket bolts. Pull sprocket off of cam nose and remove it and chain. Leave crank sprocket in place for now.

components is necessary. Nearly all chains and *cam* sprockets should be replaced—crank sprockets are relatively durable. Unless the timing set has 50,000 miles or less on it, don't even consider reusing the chain and cam sprocket.

The test for chain wear is simple: Rotate the crankshaft so the chain is pulled tight on one side and is slack on the other. Measure this slack with a straight edge. Lay it along the slack side of the chain. Push in on the chain with your thumb, midway between the two sprockets. If the chain deflects more than 1/2-in., the chain is worn-out. Replace it *and the cam sprocket.* If wear is borderline, don't waste time thinking about it—replace it anyway. Most high-mileage timing chains will all but fall off their sprockets, so you should have little trouble deciding whether or not to spend money here.

Remove Timing Chain—To remove the chain and its sprockets, undo the three bolts securing the cam sprocket to the cam. A few light pries against the back of the cam sprocket with a screwdriver will free it from the camshaft dowel.

The smaller crank sprocket can stay put. Unlike the cam sprocket, this durable, steel sprocket may not need replacing. Because it doesn't interfere with crankshaft inspection or reconditioning, it needn't be removed. To be doubly sure—with a

Nylon/aluminum cam-sprocket teeth are worn beyond reuse. If you live in a hot climate, replace it with a steel sprocket. Heat kills nylon sprocket teeth.

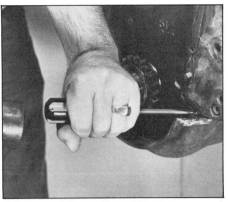

A little wedging action with a hammer and screwdriver will break the pan loose. If you remove pan with engine right-side-up, it will be easier to catch draining oil. Still, place newspapers or a drain pan underneath if you want to avoid a mess.

Oil that feels like grinding compound indicates engine may have been run without an air filter or filter was not well sealed. Poorly sealed crankcase breathers have a similar effect; the process doesn't take long if vehicle was driven on dirt roads or offroad. This is the engine with the stuck valve pictured on page 12.

silent-chain setup—drag a fingernail across the face of a sprocket tooth. If you feel deep ridges, replace it. Otherwise the sprocket is OK.

Don't hesitate to ashcan the cam sprocket; it's a weak link in the cam drive. The molded-on nylon sprocket teeth are the problem, especially in hot climates such as the Southwest desert. Heat and time combine to make the nylon teeth brittle, causing them to break off. This and a worn timing chain will cause the timing chain to jump one or more teeth. Bent pushrods are many times the unfortunate result.

Oil Pan & Pump—With the chain and cam sprocket off, turn the engine over so the pan is accessible. Now, aren't you glad you bought an engine stand? Remove the 22 oil-pan bolts and pry the pan loose with a screwdriver. Be careful not to bend the gasket flange. A lot of oil will drip out, so have a drip pan underneath.

Next, take off the oil pump. A single bolt attaches the pump to the rear main-bearing cap. Remove the bolt and rock the pump back and forth until it comes free of its locating dowel. The oil-pump drive shaft will be dangling from the pump. There's no need to separate it from the pump now, but check the plastic connector between pump and its drive shaft. It is probably cracked and brittle and should be replaced.

Connecting-Rod Numbers—Look for the connecting-rod numbers. You should find them at each bearing-cap parting line; the number on the cap corresponds to the one on the rod.

These numbers should also correspond to the cylinder number: 1 for cylinder 1, 2 for cylinder 2, and so on.

These numbers are important for two reasons: First, connecting rods and their caps are matched sets—their bearing bores are machined with the caps assembled to their rods. Only one cap will fit one rod properly. Bolt on another cap and the bore won't be round. Second, if the correct cap is installed backwards on *its* rod, the bearing bore still won't be round. So the numbers are necessary to keep the rods and caps together and in proper relationship to each other.

Your engine might have rods that aren't stamped. Maybe it is a one-in-a-thousand chance, but Chevy deals in the thousands. Also, the rods may be marked, but not in order. Number 5 could be in the number-3 cylinder. If all appears to be well—bore, piston, rod bearing and rod-bearing journal—the connecting rods should be reinstalled the same way. So, pay attention to the order—what rod is in what bore.

If the rods are out of order, note which bores they are installed in. If the rods aren't marked, mark them with a hammer and center punch or number punches. If you're using a center punch, make one punch mark for that rod number: one mark for number 1, two marks for number 2 and so on.

Ridge Reaming—It's decision time. Run your index finger up and down the top of each bore. Your fingernail will catch on a *ridge* where the top compression ring stops at TDC;

Single 7/16-14 bolt holds oil pump to block. Remove it.

Lift pump, pickup and drive shaft off of rear-main cap.

Plastic parts don't endure engine heat well. Discolored and brittle shaft-to-pump connector broke during oil-pump removal. Replace this inexpensive part, regardless of its apparent condition.

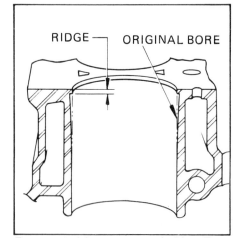

Exaggerated bore *taper* and *ridge*. Taper results from high combustion and compression pressures at the top of piston travel. Ridge is unworn section immediately above upper-travel limit of top compression ring. Drawing by Tom Monroe.

Work on one rod at a time. Position rod journal at BDC. After you've verified that the rod and cap are numbered, or you've numbered them, remove both nuts.

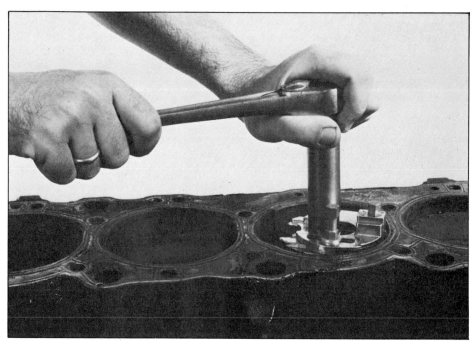

Ridge reaming. Ridge must be reamed flush with bore to avoid damaging pistons during removal. Forcing rings over ridge will damage piston-ring lands, requiring piston replacement.

where its pressure against the cylinder wall is highest. This ridge—the unworn portion of the bore—is immediately above the point of maximum bore wear. The carbon coating makes the ridge even more prominent.

If you plan on reusing the pistons, the ridges are a problem. The rings will snag on the ridges as the pistons are pushed out the top of their bores. If the ridges are small, no harm will be done. However, most engines needing a rebuild will have substantial ridges; they *will damage* the pistons.

A piston forced out the top of its bore, past the ridge, will very likely end up with damaged ring lands. However, if the pistons won't be reused because you are going to rebore the block and replace the pistons, don't worry about them. Knock the pistons out. *But,* if there's a chance that the pistons can be reused, *ream* the ridges. To do this you'll need a *ridge-reamer*—a tool specially designed to remove ridges from the tops of cylinder bores as shown in above photo.

If you don't own a ridge reamer, rent one. Follow the instructions that accompany the tool. Chances are the fellow who rented or sold it to you doesn't have an inkling as to how it should be used. So there's little point in asking him.

When shopping for a ridge reamer—whether you're renting or buying—look for one with a strong base and an adjustable cutting head. The cutting head should be sharpened only on the angled part of the blade, not the flat part. The flat part pilots the reamer in the bore. Also, get a reamer that is self-feeding—it will move up the bore as it is turned.

To fit the reamer in a bore, turn the crank so the piston moves down the bore, out of the way. Get a slide handle—T-handle—to turn the reamer. This helps prevent the cutter from cocking in the bore.

Again: **Read the directions that came with the reamer. Cut the ridge only to the worn part of the bore—no more.** If the tool has a spring-loaded, adjustable cutting head, set the depth of the cut with the spring force released. The cutter should be extended as far as the spring will push it.

If you set the cutter and then release the spring, you will cut too far into the bore. At best you will have to bore the block. At worst you may ruin the block.

Start the cut *at the bottom of the ridge,* just below the ridge's lowest point. This prevents angling the cutting head and cutting a step in the bore below the ridge.

Remove Rods & Pistons—Once you have all rods marked and ridges

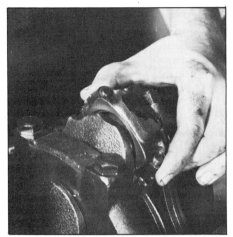

If stuck, tap side of the cap to break it loose. Use a plastic mallet if you have one. Otherwise, tap *very lightly*. Hammering will damage cap and rod. Work cap off by hand using a rocking motion.

Two short lengths of 3/8-in. hose make great rod-bolt covers. Use them to protect bearing journals and cylinder walls against nicks and scratches during piston and rod removal.

Push rod-and-piston assembly out of bore. Use a long, skinny piece of wood against bottom of the piston for this. Be ready to catch the assembly as the oil ring clears the bore.

removed, remove the rods and pistons. Start by rolling the block over so the crankcase is up. If you're working on a bench, watch your fingers! You don't want them to get flattened under that big lump of iron.

On a bench, the block will be on the deck surface of one cylinder bank while you're removing rods and pistons from the opposite bank. To allow additional clearance for rod-and-piston removal, prop up the block with a wooden block. Or position the block so the bank you're removing the piston from hangs over the edge of the bench.

Working with the block at the edge of a bench is awkward, especially when removing a rod and piston from one of the center cylinders. The block must be at the very edge of the bench. So play it safe. Have a friend hold the block on the bench as you remove the assemblies.

Turn the crank so the rod and piston you want to remove is at bottom dead center (BDC). Reinstall the damper bolt, with its heavy washer, in the end of the crank. You can now use your 3/4-in. socket and breaker bar to position the crank throws.

Remove the two rod-bolt nuts, then the cap. If the cap is tight, a few *light taps* on its side with a hammer or mallet should loosen it—don't overdo

it. If that doesn't work, tap on the end of a rod bolt with a soft mallet or the end of a wooden hammer handle. Really stubborn caps call for a different approach. Slip a 1-ft-long wooden 1x1 or broomstick past the rod and crank throw. Butt it against the piston-pin boss and give it a few light raps with a hammer. Watch the cap. Don't use your best overhand blacksmith technique, just enough impact to break the cap loose. When the rod and cap separate, grasp the cap and rock it back and forth to remove it.

Don't shove the rod and piston out until protectors are installed on the rod bolts. Otherwise the rod-bolt threads could nick and scratch the crank journal and cylinder wall. Parts stores sell special aluminum or wood "chopsticks" that do the same job, but two short lengths of 3/8-in. fuel hose are cheaper and work well.

The best way to push a piston and rod out of its bore is with a hammer handle. Grasp the hammer by the head and use the handle to push against the big end of the rod. Be ready to catch the rod-and-piston assembly when it pops out of the top of the bore. If your hammer handle is too short, use a length of wood.

Remove the protectors from the rod bolts and return the rod cap and nuts loosely—with the old

bearings—to *their* rod. Do this before you remove the next rod-and-piston assembly. Repeat this process seven more times. Start each time by repositioning the crank so the rod and piston you're going to remove next is at BDC.

Keeping rod bearings with their connecting rods will assist you later during connecting-rod and crankshaft inspection. This also applies to crankshaft main bearings. Turn to page 63 for how to *read* bearings.

Crankshaft—The crank should turn freely now that all the rods and pistons are out. If it doesn't it may be bent. Pay special attention to reading the bearings later to double-check this.

As with rod caps, the main-bearing caps should be numbered. A number should be on the machined flat near one of the main-cap bolt heads—each will be quite small. Number them in sequence yourself with a punch or numbering set if you can't find any numerals. Also look for the cast-in arrow pointing toward the front of the block. This arrow and number are necessary for proper assembly.

Those main-bearing bolts are tight; you'll need your 1/2-in.-drive breaker bar and sockets. Loosen the bolts and remove them from their caps. The bearing caps fit tightly in their registers. To remove them, use one of two methods: Tap each cap lightly on

After all rod-and-piston assemblies have been removed, the crank is next. Start by removing the main-bearing-cap bolts. You'll need a 1/2-in.-drive breaker bar for this.

Main-bearing caps *should* fit tightly in their registers. To remove each cap, give it a light tap on the side with a hammer. Keep the bearings with their caps.

its side to spring it out of the block. Or, insert the bolts back in the cap—don't thread them into the block—and wiggle the cap out using the bolts as handles.

As with the rod bearings, keep track of the main-bearing inserts. Keep them in the cap or block so you can use them while inspecting the crankshaft-bearing journals. The best way is to lay the caps upside down next to their mates in the block; then their relative positions and match-ups will be apparent. If you want to remove them, tape the insert halves together and number them 1, 2, 3, 4 or 5. You'll then know which journal they were on. As for which is the top half half or bottom half, the top has an oil hole and is grooved; the bottom half is plain.

If you find a loose-fitting cap, make a note of it. This means the cap—and probably all the others—got a real pounding. This is typical of an engine that had severe detonation. Tell your machinist so he can pay special attention to main-bearing-bore alignment. Checking bore alignment and align boring is covered in Chapter 5.

Lifting a big-block-Chevy crank out is easier said than done: It's heavy. In fact, it is best to have one person at each end. But if you are alone, you don't have much of a choice. Just be

careful that you don't scratch the main-bearing journals—lift it straight up and out.

Just in case you're wondering, the crank weighs about 70 lb. To add to the trouble of lifting is the rear oil *slinger,* or *fence.* This thin flange, immediately ahead of the rear oil seal, hangs up in the block if the crank is not lifted straight up and out. Some extra "umph" at the rear of the crank should bring it straight out, avoiding any hang-up.

Remove Camshaft & Lifters—The remaining moving parts in the block are the camshaft and lifters. The trick is to remove them without damaging them or getting the lifters, or tappets, out of order—very important if you want to save both the cam and its lifters.

A used cam lobe and its lifter is a matched set; they must be kept together. Although a new lifter can be installed on a used cam lobe, you'll have to replace both cam and lifters if the cam is replaced. And the used lifters must be kept in order—I discuss the reasons for this on page 80. Otherwise the cam lobes and lifters will be *wiped,* or worn down to nothing in only a few minutes of engine operation due to excessively high lobe-to-lifter pressure.

However, if you think the cam has a

bad cam lobe, you don't have to keep the lifters in order: If you have any doubts, look at the cam. You can see it clearly now that the crank and rods are out of the way. Although further damage to the cam is not a concern, be careful of the cam bearings. You may want to save them.

Before the cam can be removed, all lifters must be raised off the cam lobes and held there. Raising the lifters can be done simply by turning the cam at least one revolution. The problem, then, is to keep the lifters raised in their bores. Solve this little dilemma by either setting the block up on its rear face or rolling it over on its back—accumulated varnish around the foot of each lifter will keep them from falling out.

Loosely fit the cam sprocket to the nose of the cam, engaging it with the dowel. Turn the cam one complete revolution to move the lifters out of the way. Remove the sprocket.

Note: A gear-driven cam is retained by a thrust plate This plate is held to the front of the block with two bolts. These bolts can be reached with a socket and extension through the holes in the cam gear. Once the thrust plate is loose, the cam can be pulled from the block using the same procedure as with the chain-and-sprocket-driven cam.

If you can pull valve lifters out of their bores, remove and store them in order. This is a must if you plan to reuse cam and lifters. If lifters won't come out, use cam sprocket as a handle to turn cam one revolution. This will raise the lifters in their bores so you can pull out cam. Block should be on its side or upside down when you do this.

Lifting crank out is a real backbreaker. It must be lifted straight up out of block. This job is much easier with one person at each end of crank. Leave main bearings in place so you can *read* them later.

Thread a long bolt into the chain-driven cam nose so you'll have something to pull with. Sometimes it's hard to start the cam moving. If pulling on it doesn't work, use a screwdriver to pry against the backside of a cam lobe or bearing journal. Don't let the screwdriver slip and scratch a lobe or journal.

With one hand inside the block to guide the cam and the other on the nose, pull the cam out. Be careful to keep the cam lobes from hitting the bearings. The hard lobes will damage the soft camshaft bearings. You want to save the bearings, if possible.

Don't move the block. The lifters are now free to fall out the bottom of their bores.

Lifters—Before you push the lifters out, make provisions to *keep them in order*. Egg cartons work, but a 2x4 with sixteen 7/8-in.-diameter holes drilled part way through in two parallel rows of eight is best. Make sure that whatever you use is permanently marked FRONT to represent the front of the engine. Place the lifters in the container in the order in which they came out, starting with the front.

As you push a lifter out the bottom of its bore, keep your hand underneath to catch it as it falls out. You don't want it to bounce around inside the block or fall on the floor—unless

Using two hands, be very careful when guiding cam out of the block. Cam bearings are easily damaged. Additionally, a cam lobe or bearing journal can be damaged by bumping it against the block. Before you roll the block back over, pull lifters out of the bottom of their bores and store them in order.

you know *positively* that you aren't going to reuse the lifters.

Gallery Plugs & Cam Bearings— Don't put your tools away; there are still a few items that need attention in the block.

One of the best features of the big-block Chevy is its bullet-proof oiling system. If your engine has been well

cared for—with frequent oil and filter changes—and you found no sludge inside the engine, you don't have to clean the oil passages. You can't get them much cleaner than they are! On the other hand, if the engine looks as dirty as the one pictured, remove the oil-gallery plugs so you or the machine shop can clean the oil passages.

Access to the oil galleries can be obtained by removing the threaded gallery plugs. There are 10 of them; nine Allen plugs and one square-head plug. Two Allen plugs are in the front of the block, adjacent to the number-1 cam-bearing bore. There's another in the front of the block, at the extreme bottom left. Two are at the rear of the block, next to the cam bearing and in line with the two front plugs. You'll find the square-head plug back there, too.

A row of four gallery plugs stretches along the bottom left side of the block, directly opposite the front four main-bearing webs. Each of these plugged galleries leads to a main bearing and cam bearing. The rear main- and cam-bearing passage is accessed through the oil-pressure sending-unit hole. High-performance blocks have two additional tapped holes with plugs above the oil-filter adapter. These are for plumbing in a remote oil cooler. Leave them in place unless the block is really filthy.

Before trying to loosen the oil-gallery plugs, clean out the hex recess in the Allen plugs. Use a pick, scribe or small nail to dig out the grime. Residue left here will prevent full engagement of the Allen wrench. Removing a stripped Allen plug is very difficult—work to avoid it. A sloppy-fitting wrench will have similar results; the Allen wrench should fit snugly.

If the end of the wrench is worn, hacksaw it off to obtain a fresh hex. Remove the sharp edges and burrs with a grinder. If you have an impact wrench like the one pictured, use it. A couple of wallops with a hand impacter will break a gallery plug loose. Without one, you're headed for a stripped plug.

Cam Bearings—Now is the time to consider camshaft bearings because they make a difference in how you clean the block. You may have detected a few light scratches on the cam bearings, but that doesn't mean they have to be replaced.

As a rule, cam bearings don't wear out. They are lightly loaded and drenched in oil. Only when crankcase oil remains extremely dirty will the bearings be worn to the point where they need replacing. Therefore, the chance of the cam bearings needing replacement is *very* unlikely.

That's good news because: Chang-

Oil-gallery plugs are difficult to remove. A sharp impact from an impact driver, such as this, and a hammer is what's needed to get an oil-gallery plug moving. Regardless of what you use, dig all deposits from the heads of the plugs. This will give the Allen wrench maximum engagement.

ing cam bearings is a job for the machine shop. A special *mandrel* and drive bar or pull bar is needed to remove *and* install cam bearings, page 65. It isn't a job for the amateur. This eliminates a chance to save some money if the cam bearings need to be replaced. If they don't, you won't have to spend any money in this department.

All this takes on more meaning when it comes to cleaning the block. Two commercial methods are used to remove grease and scale from an engine block: *jet-spraying* and *hot-tanking*. If your engine is clean enough and you decide not to open the oil galleries for cleaning, then jet-spraying is for you. A high-pressure stream of solvent is blasted on the block, washing away grease and dirt. *Non-caustic* solvent—the kind frequently used in a jet-sprayer—doesn't "eat," or dissolve bearing material.

The one disadvantage of jet-spraying is the inability to clean inside the oil galleries and water jackets. This will have to be done with degreaser, gun-bore brushes and elbow grease.

On the other hand, the solution used in hot-tanking is caustic. Any non-ferrous part, such as aluminum, lead or plastic, will dissolve in a hot-tank. So, although a hot-tank will do the best job of cleaning your block, it will destroy the cam bearings. Some core plugs won't survive a trip through the hot-tank either, but you need to replace them anyway.

If you do have your block hot-tanked, plan on replacing the cam bearings; they won't be in the block

At left is aftermarket spin-on oil-filter adapter. At right is original cartridge-type adapter. If you have the cartridge-type filter, retain it. Big-block cartridge filter holds more oil and costs less.

when it's pulled from the tank! Nor will the employee of the machine shop. The dissolved bearing material will dilute the expensive hot-tank solution, rendering it useless. Therefore, the bearings must be removed before the block is *tanked*.

If you don't know which way to go, opt for the hot-tank. This means you'll have to pay for new cam bearings, plus their removal and installation, but hot-tanking is the *best* block-cleaning method.

Oil-Filter Adapter—In the block's oil-filter recess you will find the oil-filter adapter. It must be removed for cleaning and inspection. Because it is aluminum, it must be removed before the block goes into the hot-tank.

Oil-filter adapters used on '65—'67 engines are for cartridge-type oil filters; '68 and later adapters use spin-on filters. Regardless of the type, remove the two bolts near the adapter center. You shouldn't have to do much more than lift out the adapter. If it's stuck, pop it loose with a screwdriver.

If you have a cartridge filter, think twice about converting to a spin-on. Not only will the adapter cost several dollars, you'll find the cartridge-type filter is cheaper. It also has more filtering area. Stay with the cartridge-type filter.

Although the filter adapter was changed, all '68 and later truck and police engines were equipped with cartridge filters, even after passenger-car engines were equipped with spin-on filters. I just gave the reasons for this; greater filtering area and lower cost were considered more important than the spin-on's convenience.

If your engine has a spin-on filter

Main caps with their insert halves in position for _reading_ bearings.

and you want to switch to a cartridge-type filter, you can use the '68 and up truck/police filter adapter on any '68 and up big-block.

BEARING INSPECTION

All big-block-Chevy bearings—rod, main and cam—are the _precision insert_ type. They are removable shells, not poured permanently into the block, then sized afterwards. A _trimetal_ bearing, lead-based _babbitt_ is poured over a _substrate,_ or _steel shell._ On top of the babbitt there's a thin layer of tin. Except for the rear-main bearing, the first four main bearings are the same. The rear main bearing is flanged. It resists crankshaft thrust, not just radial loads.

Except for the top half of the front main bearing, you will probably find most of the bearings are not worn through the tin overlay. If your engine is loaded with belt-driven accessories, such as an A/C compressor, air pump and power-steering pump, the constant upward load of the drive belts on the crank nose and bearing causes more rapid wear on the number-1 top bearing half. With the lead-babbitt bearings used in big-blocks, this shows up as a darker shade of gray.

Unlike lead-babbitt bearings, copper-babbitt bearing material contrasts sharply with the tin overlay. A copper-babbitt bearing shows up clearly as copper color against the gray overlay. In the big Chevy you have to keep your eye peeled for the underlayer of babbitt that's just a little bit darker than the overlay. If your garage is dark—most are—and you're having trouble seeing, use a flashlight, droplight or sunlight.

Other than the uneven wear just discussed, uneven wear from front to back around the full circumference of a bearing is abnormal. This means the bearing journal is tapered—larger in diameter at one end than the other. Although this will show up when you measure the main-bearing journals, this bearing-wear pattern is a warning sign. Also, if the thrust bearing is badly worn it will allow the crank to _walk_ forward. The main-journal _fillets_—radius at the ends of the journals—will run into the main bearings. This exposes the thin line of babbitt around the bearing edge.

Scratches indicate dirty oil resulting from faulty air and oil filters. There was an update on very early big-block bearings to make them more _embeddable._ The bearings were sacrificed in favor of the bearing journals.

When a hard foreign particle gets trapped between the bearing and journal, it _embeds,_ or gets pushed into the bearing, rather than scratch or groove the journal. Check the bearing for embedded dirt. Also check the crank journal of a scratched or dirt-embedded bearing. It may also be scratched or grooved.

Worse than dirty lubrication is _no lubrication._ If your engine has ever lost lubrication, the main and rod bearings will be _wiped_—worn to the babbitt, or possibly through it, over the entire bearing surface. Depending on how long the engine was without lubrication, the journal may be heavily scored. Only one bearing will be wiped if the oil passage to that journal was clogged. However if all the bearing journals are damaged, the oil pump failed or crankcase oil-level was too low. Triple-check the oil-pump clearances and pressure-relief spring using the procedure detailed in Chapter 5.

After you've examined each bearing closely, arrange the caps upside-down on the block so you can see all the bearing surfaces at one time. Stand back and compare the conditions of the bearings. If just one bearing is wiped badly, _rod-out_ its oil passage—run a welding rod or straightened coat hanger through. Also check the two connecting rods fed by that main; they are probably wiped badly too. If the main bearings get progressively worse the farther they are from the rear main bearing, then either the oil pump or oil level could have caused the problem. The oil system feeds from rear to front, so if oil is insufficient, the front bearings are first to suffer.

Most main-bearing problems are oil related, but a bent crank will also cause problems. If so, your common sense is your best friend when reading the patterns of worn-out journals. You'll need common sense because bearing patterns are unpredictable when caused by a bent crank. Depending on how long the bend is, you may find two, three or more bearings affected.

Usually, you'll be able to see effects of the bend over at least three bearings. Just be on the lookout for a bearing wiped out entirely with its neighboring bearings worn out on the top and bottom on opposite sides.

If bearing-3 was the high point of the bend, for example, it will be worn out around its entire circumference. The bearings next door, bearings 2 and 4, will be wiped out too, but with a distinctive pattern. The upper half of bearing 2 will be worn, but only the side away from bearing 3. The lower half of 2 will be worn also, but on the side closest to bearing 3. Bearing 4 will be worn in the same way as bearing 2.

5 Shortblock Reconditioning

Checking bore diameter with inside mike before making final pass with boring bar. Truck big-blocks can be safely overbored 0.125 in. Before you do it, check piston availability.

Let's start the decision-making stage of the rebuild by measuring and reconditioning the parts that make up the shortblock. The foundation for your new engine is the shortblock. So it's important to understand that mistakes made here probably won't show up until the engine is back in service. And they'll be expensive to correct.

Usually mistakes are made when

there's not enough information available or because of carelessness. I'll provide you with all the information you will need to rebuild your engine correctly. It's up to you to provide the care. Use your common sense. If done correctly, your efforts will be rewarded with an engine that runs as well or better than it did fresh from the factory. Skipping steps may save

time now, but will cost you later in terms of engine longevity or costly, time-consuming repairs. Don't shortcut.

CLEAN & INSPECT CYLINDER BLOCK

What can ultimately be determined from inspecting the head gaskets and gasket surfaces is the flatness of the heads and block. As the owner of a big-block Chevy, you know durability is a major benefit of this engine's size and weight. So, when it comes to warped engine parts, about the only ones you may find will be the heads.

The massive heads and number of head bolts mean you *probably* won't find any problems with yours. Regardless, study the head gaskets and the patterns left by them on the block and heads. Look for signs of combustion or water leaks.

Combustion leaks show up as dark streaks radiating from the combustion chamber or cylinder. The leak may lead to another combustion chamber or cylinder, the water jacket or the atmosphere.

Called *notching,* a combustion leak between cylinders can erode away block material if the engine is operated under load for a sustained time. Combustion gasses act like an acetylene torch to cut, or notch, the block. If your engine had such a leak, it would have been obvious if you did a power-balance test, compression or leak-down test. If you didn't perform any of these tests, think back. How did your engine run? Rough idle and low power point to a leaky gasket.

As for leaks into the water jacket, you should recall additional symptoms: rapid engine warmup, coolant loss and overheating. The gasses would show as bubbles in the radiator or during a sniff test. This type of leak shows as rust streaks combined with dark streaks on the block or cylinder-head gasket surfaces.

Combustion leaks to the atmosphere cause a spitting sound, and will show as black streaks running from a cylinder to the edge of the block deck or cylinder-head gasket surface. If the leak also got into the water jacket, look for rust streaks down the side of the block.

As you inspect the gaskets, keep in mind that dark areas represent a possible leak. Normal usage will cause some discoloration. If the combustion or coolant discoloration is small—under 1/4-in. square—don't worry about it.

A new head gasket properly installed will seal. It's the big leaks you should be concerned with. Look for heavy streaking marks where lots of water or gasses have escaped. If you find any signs of head-gasket leakage, check both block and cylinder-head gasket surfaces for imperfections and warpage, pages 71 and 97.

Gasket Scraping—Remove all of the old gaskets with a gasket scraper, putty knife or single-edge razor blade. While you do this, continue to look for signs of leakage.

If steel-shim gaskets were used, you'll have an easy time cleaning the head and block. Shim-type gaskets are

Antifreeze should be used year-round. Here is one result of using pure water for an extended time period; heavy deposits in the water jacket. A block that is this dirty should be hot-tanked.

installed without sealer, so there will be little residue once the gaskets are pulled off.

Composition gaskets are another story; gasket material sticks to both block and head. It must be scraped off. If a valve job or other internal

engine work has been performed on your Chevy, you'll probably find a composition gasket stuck on with sealer. This really leaves a mess, so you'll be doing some heavy gasket scraping.

Of the three gasket-removal tools I mentioned, the gasket scraper is the best. It looks like a screwdriver with a wide chisel-type blade. It will save you lots of work, and is much safer than pushing a razor blade or a flexing putty knife.

As long as you have the scraper in hand, clean the front cover and oil pan. Look for damage to the sealing surfaces. Sight down the flanges of the oil pan and timing cover; these sheet-metal items are easily warped or bent. Look also for dimpled bolt holes. This occurs if the bolts are overtightened. Straighten and flatten a bent or dimpled bolt flange with a hammer on an anvil or other flat, firm metal surface.

Dutch Cleaning—There are several ways to clean a block, as discussed in Chapter 4. Actually the block must be

Before a block goes into the hot-tank, the cam bearings must be driven out.

cleaned three times: Once so you or your machinist can take measurements to determine what work must be done, and two more times after all machining is complete. In the next couple of paragraphs I'll describe various cleaning methods in detail and let

Gasket scraping is much easier after block has been hot-tanked. Use a gasket scraper or sturdy paint scraper to save your knuckles.

you decide which one is for you.

Do your first cleaning with a *canned* degreaser and the garden hose. The initial measuring and inspecting can then be done. The second cleaning is done at the machine shop in the hot-tank or spray cleaner. With the hot-tank, parts are boiled in caustic cleaning solution. Spray-type cleaners spray the parts with solvent under high pressure. You'll do the final cleaning at home with soap, water and a scrub brush.

If you've decided to replace the cam bearings, you can either hot-tank or spray clean your block. I think the hot-tank is the best way to go.

The hot-tank—unlike spray cleaning—gets *inside* the oil and water passages—if you removed *all* the oil-gallery plugs. In addition to the block, tank *all iron and steel* parts with the block. These parts include iron heads, crankshaft, camshaft, nuts and bolts, iron intake manifold, tin work and rods. Do not hot-tank any die-cast metal, plastic or aluminum parts, otherwise you won't see them again.

The longer the parts are in the tank, the cleaner they will get. Massive parts, such as the block and heads, will take more time to clean than light pieces. Your machinist knows how much time each will take.

If you want to save the cam bearings, find a machine shop with a jet-sprayer and non-caustic solvents. Volkswagen repair shops are a good place to start. The large amounts of nonferrous parts in air-cooled VW en-

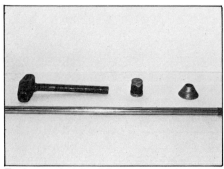

Tools needed to drive out cam bearings: drive bar, hammer, mandrel and centering cone.

After hot-tanking block, caustic solution is sprayed off with steam cleaner. Special attention must be paid to oil galleries and water jackets.

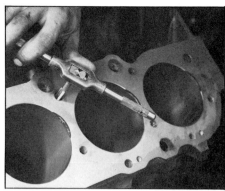

Chasing threads with a bottoming tap. This necessary step will remove dirt and corrosion to ensure more accurate head- and main-bearing-bolt torques. After running tap in an out of each hole, blast out dirt with compressed air.

gines require such a cleaner. Of course you could do the job yourself, especially if you are on a tight budget. Do-it-yourself car washes with their high-pressure sprayers work very well. And there is no problem with the run-off.

Follow-up at home with a can of engine degreaser and the garden hose. You still have to get into oil and water passages. This can be done with a small rag tied to a wire, pipe cleaners, a round file and rifle-bore brushes. Both 0.30- and 0.32-caliber brushes work well. Carburetor cleaner will loosen the tough stuff, but wear eye protection and rubber gloves. Don't splash this strong solvent around. Keep pets and kids away from it too.

Threads—Another cleaning job involves cleaning the threads by *chasing* them. This is done by running a tap into the bolt holes. Most important are the head- and main-cap-bolt threads, but all the rest need a clean-up too.

You need *bottoming taps,* as opposed to *taper* taps or *plug taps* for threaded holes in the block or heads. A bottoming tap is fully threaded, all the way to its end. Unlike taper or plug taps, which are tapered at their ends, a bottoming tap cleans the very last thread in a blind hole.

It's surprising how much dirt and grime a tap will drag out of a bolt hole. It is important to remove all this junk, otherwise the bolts cannot be torqued properly. Dirty threads cause binding, resulting in less tension being applied to a bolt for a given amount of torque.

Proper torque is very important with head bolts, especially if inconsis-

tencies result from dirty threads. Not only will there be insufficient bolt tension, tension will be uneven from one bolt to the next. This increases the chance of a blown gasket. An improperly torqued main-bearing bolt may cause a spun bearing.

Use a *7/16-14* tap—7/16-in. diameter, 14 threads per inch—for the head-bolt threads. A 1/2-13 tap is used for the main-bearing caps. Clean all bolt threads, too. Use the same-size dies, from your tap-and-die set, or use a wire brush. A wire-wheel brush on a bench grinder does an excellent job of cleaning bolt threads. Remember to wear eye protection.

Smart mechanics always *buy* the best tools. Smart *poor* mechanics always *borrow* the best tools. This should hold true when buying or borrowing a tap-and-die set. A tap or die is designed to cut threads; cleaning them is secondary. If they are of poor quality they might have a slight mismatch with the threads being chased. For instance, if a tap is too large, it will cut the existing threads. This will damage or weaken the threads.

Most low-quality tap-and-die sets are imported. But that doesn't mean all imported sets are bad, of course. Stay with the big-name tool makers and beware of ''bargain'' tap-and-die sets.

A real bonus during the cleaning process is a source of compressed air and a blowgun. You'll then be able to blow dirt and water out of hard-to-reach areas and air-dry washed parts. Be careful that dirt or water doesn't ricochet back into your face or onto bystanders. To be on the safe

side, wear goggles or a face shield when using a blowgun.

Air drying is a real help. Machined engine parts, such as cylinders, bearing bores and bearing journals, rust very quickly after washing if they are not dried. This is especially true when soap and water are used. Compressed air is the quickest way to dry a part.

Immediately after rinsing off soapy water with clean water, dry the block. If you don't have compressed air, use paper towels. Just be quick about it.

Follow the drying with a protective coat of *water-dispersant* oil. CRC and WD-40—available in spray cans—are popular examples. Rather than trap moisture underneath—as does motor oil—water-dispersant oil gets between the moisture and the metal surfaces to prevent rust. Additionally, spray cans are more convenient to use than a squirt can of oil.

Use motor oil if that's all you have. It's better than leaving freshly cleaned machined surfaces unprotected. Just make sure the surfaces are totally dry before you oil them. The major areas to be concerned with on the cylinder block are the cylinder bores and the cam- and crankshaft-bearing bores.

FINAL BLOCK INSPECTION & MACHINING

After you've cleaned the block, determine what is required to put it back into top shape. To inspect the block and bottom-end components, you will need a few specialized tools.

To measure the pistons and cylinders you'll need a 3—4-in. or 4—5-in. outside micrometer and a 3—4-in. or

Drying block with compressed air. This is great for drying machined surfaces and blowing out oil galleries and threaded holes. Follow drying by spraying machined surfaces with water-dispersant oil.

Dressing sharp edges with a smooth, flat file ensures against small nicks or high-spot. This will ensure against the block being positioned incorrectly in the boring bar or mill. Light passes are used on all corners. Don't lean on the file or you will remove too much material.

4—5-in. inside micrometer or telescoping gage, sometimes referred to as a *snap gage.* A 2—3-in. outside micrometer is needed for the crankshaft rod- and main-bearing journals. The 3—4-in. mikes are used to measure 366 and 396 bores; 4—5-in. mikes are for the others. If your engine had a leaky head gasket, check the heads and block decks for flatness. This calls for a precision straightedge and a set of feeler gages. Most tools can be rented—or pay the machine shop to do the measuring.

Checking Bore Wear—How much cylinder bores are worn *always* determines whether the block should be bored or merely honed. Boring automatically means new pistons. Otherwise the bores can be honed, or their *glaze broken,* page 71. You'll be able to reuse the old pistons, *provided they are reusable.* Read the piston section, beginning on page 83, before making your final decision.

There are three ways to measure bore wear. The most accurate and convenient is with a *dial bore gage.* But this tool can only be found in an engine machine shop. Next best is with the inside micrometer or a telescope gage and an outside micrometer. The last method is to use an old piston ring and a set of feeler gages. Ring *end-gap* change from the top to the bottom of a bore is directly related to bore wear, or *taper.*

Bore Taper—Remember the ridges at the tops of the bores? Notice how the bores are worn at the top of the piston stroke, but not at the bottom. There is no bottom ridge! The reason is simple. Bores don't wear evenly from top to bottom. Look at the cross-section drawing of the cylinder. Notice the bell-mouth, or taper, the bore wears into. This is because piston rings are forced against a cylinder bore mostly by compression and combustion pressures, rather than ring tension.

Compression and combustion pressures are greatest at TDC, so the force exerted by the rings against the bore is greatest at the top of the piston stroke. Therefore the bore wears more at the top than the bottom. In fact, a bore wears very little midway down—so little that the original *cross-hatch* hone pattern is probably visible at the lower end of each bore.

Measuring Taper—Measuring bore taper involves finding the difference in diameter between the unworn portion of a cylinder and that of the greatest wear. You need to note these measurements for each bore; not all cylinders wear the same amount or in the same places. Start by taking measurements parallel to the block centerline and 90° to the centerline. Then take measurements in between to establish maximum wear.

Maximum bore wear is *close* to what it will take to *clean up* a cylinder—or how much has to be re-

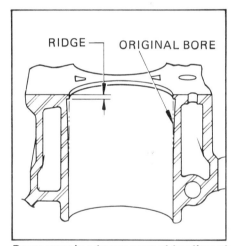

Bores wear in a taper: more at top than at bottom. Short unworn section at top of bore is ridge, directly above top-compression-ring-travel limit. Drawing by Tom Monroe.

moved from that cylinder so new metal is exposed from top to bottom. Measure all cylinders. The one with the greatest wear governs how much will be removed from the remaining seven. If you find one cylinder that's badly damaged to the point that it can't be corrected by boring, consider *sleeving* it. Read the sidebar, page 68.

You may notice that *front* and *rear* bores show greater wear than others. The reason is interesting; the warmer the cylinder gets, the less it wears. And the end cylinders run cooler that the center ones. Additionally, the front wall of the front cylinders and

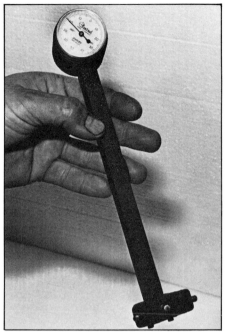

Dial bore gage is operated by plunger at end opposite dial. Gage is inserted down into bore, dial is zeroed, then gage is moved to top of bore, immediately below the ridge. Bore wear is read directly.

the rear wall of the rear cylinders run even cooler, particularly during engine warmup. This causes the cylinder to become egg-shaped. You'll be able to "feel the difference" in bore wear if your engine has prominent ridges. Run your fingernail around them.

The final factor in measuring bore wear is the uneven taper wear pattern we've already seen. Because wear is not concentric—it is uneven around the circumference of the bore— maximum-taper measurement will not give you the final overbore figure. That's because the non-concentric

bore wear moves, or *shifts,* the bore center. So you will always have to bore a little more than the taper-wear figure indicates.

Your machinist will begin boring at the worst cylinder. Many shops start with a 0.030-in. overbore because of piston-size availability. This *almost* en-

Sleeve is simply a replaceable cylinder. If one cylinder can't be reconditioned by boring, consider having it sleeved. If more than one requires sleeving, consider replacing the block. What you should pay for reconditioning depends on value—or rarity—of the block. Photo by Tom Monroe.

SLEEVING AN ENGINE
Blocks with nothing wrong except for one cylinder that's not *serviceable*—cracked or very deeply grooved—can be saved by *sleeving.* A sleeve is a replaceable cylinder, similar to a section of pipe.

To install a sleeve, the damaged cylinder is removed by boring. The new cylinder, or sleeve, is installed and bored and honed to match the others, standard or oversize. Machine shops charge about $50 to sleeve a cylinder—a lot less than the cost of a new or used block.

A cast-iron sleeve is slightly longer than the original cylinder and its bore is smaller than the standard bore. The smaller bore allows the machinist to finish the cylinder to the desired size, standard or oversize. Sleeve walls are 3/32—1/8-in. thick.

First step in a sleeve installation is boring the damaged cylinder about 0.001-in. smaller than the sleeve OD, giving an interference fit. Additionally, the cylinder is not bored all the way to the bottom of the block, but is stopped short, leaving a shoulder at the bottom.

The ledge and interference fit combine to position and lock the sleeve in the block so it will not move. A common error in sleeve installation is to provide more than a 0.001-in. interference fit. This is not

good because the neighboring cylinders will be distorted. I'm not suggesting that you tell your machinist how to do his job, but working with an established shop will avoid such errors.

After boring, some shops heat the block with a torch or furnace to make it expand. At the same time the sleeve is cooled in a freezer so it will contract. This momentarily eliminates the interference fit and permits the sleeve to drop into the block—most of the way. Temperatures of the block and sleeve converge rapidly, however, and the sleeve installation has to be completed by driving it into place. Before the sleeve is driven all the way "home," it is sealed around its bottom edge.

Some machine shops don't bother with the heating and cooling process, preferring to press the sleeve all the way in. Regardless of how it's installed, the portion projecting out the top of the block is trimmed flush with the deck surface. Boring and honing is the final step.

Position ring immediately below ridge and measure end gap with feeler gages. Record measurement. Push ring to bottom of bore and remeasure end gap. Subtract end-gap measurements and refer to the nearby chart to determine bore wear, or taper.

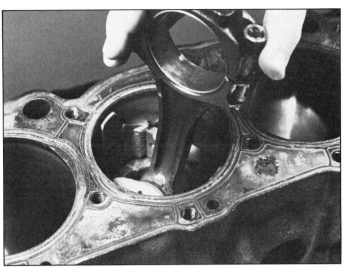

To get accurate end-gap readings, ring must be square in bore. Use *flat-top* piston to push ring down bore to square it up in the bore. Pop-up pistons won't work.

sures that all cylinders will come clean without the need to bore twice. Sometimes engine machine shops don't even check wear other than running a fingernail up and down the bores, and feeling the ridge. If there isn't anything apparently wrong with the cylinders they'll just set up the block in the boring machine and make eight 0.028-in. passes.

Bore Gage or Micrometer—When checking bore wear directly, with either a dial bore gage or telescoping gage and/or micrometer, first measure right below the ridge. Measure at different points around the bore to find maximum wear. Now, measure near the bottom of the bore, where little wear occurs. Find the greatest bottom dimension. Subtract this figure from the maximum figure you got at the top, immediately below the ridge. This figure is bore taper. Write down each measurement as you make it.

Ring and Feeler Gage—Unlike reading taper directly with a dial indicator, using a piston ring and feeler gages is an *indirect method* of measuring bore wear. You are comparing the *circumferences* of a cylinder at two points, not its *diameters,* so this method doesn't register irregularities in bore wear. This makes it less accurate than direct-measuring methods, but it will tell you if you'll have to bore or not.

To measure taper by the ring/feeler-gage method, place a compression ring in the bore. Square it up in the bore with a piston turned

upside down and without its rings. You can't use a domed piston. If your engine has them, you'll have to find a flat-top piston or a tin can that fits the bore snugly. Or simply use a 6-in. scale and measure from the block deck to the ring to square it.

Using feeler gages, check ring *end gap*—the space between the ring ends. Take two measurements in each cylinder: one immediately below the ridge, another right at the bottom of the bore. Do all eight cylinders, using the same ring. Record your readings as you go, then use the chart, table or formula to determine taper. Taper is approximately the difference between top and bottom ring end-gap measurements *multiplied by 0.30.*

Allowable Taper—What you establish as *maximum allowable taper* depends on the durability you expect from your engine after it's back in service. If you want nothing more than a quick fix that probably won't go more than 10,000 miles before it starts burning oil and is down on compression, then you can overlook anything up to 0.010-in. of taper. But, if you want a 100,000-mile big-block, 0.005-in. taper is maximum. Any more and the engine must be bored. Those figures are for direct measurements. If you use the ring and feeler-gage method, the numbers are less at 0.008-in. and 0.003-in., respectively.

If you decide not to bore, remember that taper will only get worse. And the ring durability goes down with in-

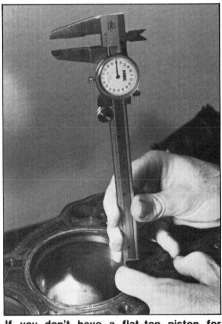

If you don't have a flat-top piston for squaring the ring, use a 6-inch machinist's rule, the depth-gage end of a vernier caliper or whatever to gage from the block-deck surface to the ring.

creased bore taper. Rings expand and contract on each trip up and down a tapered bore. Therefore, as taper increases, so does ring flexing; and they fatigue sooner. Piston ring lands also take a beating. As the rings flex, they rub against the ring lands, accelerating wear and reducing sealing. Therefore, if the 0.005-in. direct measurement or 0.003-in. indirect measurement is exceeded, your best move is to bore the block and install new pistons.

$G_2 - G_1$ ΔG	TAPER
0.000	0.0000
0.001	0.0003
0.005	0.0016
0.010	0.0032
0.015	0.0048
0.020	0.0064
0.025	0.0080
0.030	0.0095
0.035	0.0111
0.040	0.0127
0.045	0.0143
0.050	0.0159
Approximate Taper = 0.30 X ΔG	

Taper is found by subtracting the two end-gap measurements, or $G^2 - G^1 = \Delta G$. Use the the curve or chart to find bore taper. Or simply multiply ΔG by 0.30.

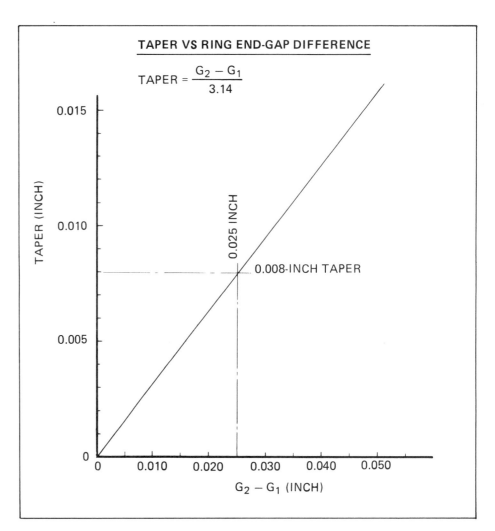

TAPER VS RING END-GAP DIFFERENCE

$$TAPER = \frac{G_2 - G_1}{3.14}$$

0.025 INCH

0.008-INCH TAPER

TAPER (INCH)

$G_2 - G_1$ (INCH)

Measuring piston-to-bore clearance directly with feeler gages. Check each piston in *its* bore. When clearance exceeds 0.0025 in. with a cast piston, junk it. Forged-piston clearance can go up to 0.0055 in. or more, depending on the engine.

Piston-to-Bore Clearance—Before you make the final decision regarding taper and the need to bore, there is another determining factor; *piston-to-bore clearance*. After all, if the pistons are worn out and must be replaced, rebore. The cost of new pistons is the major expense, not the machine work. So rebore and start out with straight bores, even if the taper is minimal.

An exception might be made where one piston needs to be replaced and there is little bore wear, or taper. You might be better off replacing that piston, even though it should be balanced to match the others.

There are two ways to measure piston-to-bore clearance: *direct* and *indirect*. In this case the direct method is easier and cheaper than the indirect method.

To measure directly, select a piston without rings and insert it upside down in *its* bore. Rotate the piston so its *front* points to the front. The "dot" or arrow on the top of the piston goes to the front of the block. It doesn't matter if the piston is upside down—this is the easy way to hold it in the bore.

Using feeler gages, find the maximum distance between the bore and the piston *thrust face*. The thrust face is the piston's right skirt—left looking from the bottom with the piston pointing forward. Take several measurements with the piston at the top and bottom of its travel in the bore, between TDC and BDC. Record your findings.

Indirect measurement requires 3—4- or 4—5-in. outside micrometers to measure the pistons. The bores must be measured with inside mikes or a telescoping gage and the outside mikes. The difference between the two is piston-to-bore clearance.

Using the outside mikes, measure piston diameter across its skirts. Write down this measurement. Now, take an *inside mike* or a telescoping gage and measure the cylinder bore 90° to the piston-pin centerline about 2-in. down the bore. Record this figure. Now subtract the two measurements to find piston-to-bore clearance.

With either method, make sure *both* piston and block are the same temperature. If the block is outside in the sun and the pistons are in a cold garage, your measurements could be off as much as 0.001-in.

How Much is Too Much?—Well, it depends. Cast pistons need less assembly clearance than the forged variety. For instance, Chevy recommends 0.0007—0.0015-in. clearance for cast pistons, but with a 0.0025-in. wear limit. So if piston-to-bore clearance is close to or over 0.0025 in., it's replacement time.

Forged pistons are another story.

For starters, they are fitted with a larger clearance—somewhere between 0.0036 and 0.0063 in. depending on the HP rating of the engine. The higher the horsepower, the greater the clearance. Wear limits of forged pistons vary also, from 0.0055 to 0.0085 in. Again, the higher the HP, the greater the wear limit.

You need to remember the long list of Super Heavy Duty parts Chevy has made for the big-block over the years. These *off-road* parts are included in the forged piston-to-bore clearances and wear limits. The rule when it comes to a post-'67 big-block Chevy states that forged pistons are found in engines with mechanical camshafts. No mechanical cam, no forged pistons. Also, the "hotter" the cam, the more piston-to-bore clearance.

There's no question as to the need to replace this piston. Preignition caused piston to overheat, scuff, then seize in bore. Photo by Steve Christ.

To sum it up, if you have an LS-6 cam, forged-piston wear limit is lowest at 0.0055-in. bore clearance. If you are lucky enough to own an L-88, it uses a slightly higher wear limit, 0.006-in. I recommend that you use the lower LS-6 limit, or 0.0055 in. Deciding on clearances also depends on what you want to do with your engine. If you are restoring an engine for normal service, use the lowest stock clearances with new pistons to minimize oil consumption and piston

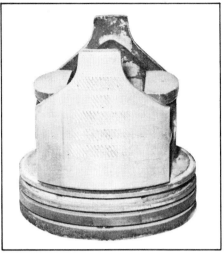

Embossed patterns rolled into piston skirt are *knurls*—a temporary fix for reducing piston-to-bore clearance. It won't take long for the knurls to wear off, putting piston-to-bore clearance back where it was before the knurling.

noise, and give it the longest possible piston and bore life.

If you are "going to the races," wear limits should not be a factor; they shouldn't even be considered. First, you should start by installing new pistons in fresh bores. And, next, piston-to-bore clearances should be set at the high end to allow room for additional piston expansion and to reduce internal friction at the higher operating temperatures encountered under racing conditions. For a street-driven engine, stick with the tighter clearances; you'll enjoy this kind of engine a lot better than an oil-burning, rattle-in-the-morning loose runner.

Another fact to consider: Chevy and TRW forged replacement pistons feature a slight *barrel,* or oval shape above the pin. This allows them to be fitted more tightly in their bores than replacement pistons from other sources. If your engine has had other forged replacement pistons installed, replace them with Chevy's or TRW's for better oil control and less noise. Your engine will get the same horsepower, or more. Regardless of the pistons you install, *use the manufacturer's recommended clearances.*

Piston Knurling—Don't be talked into *knurling* your pistons to take up piston-to-bore clearance. Knurling involves *embossing,* or rolling a pattern into the piston skirts. This creates indentations and displaces material to form tiny ridges on the skirt. Excessive piston-to-bore clearance is reduced *at the top of the ridges only.*

If original pistons and bores are to be reused, the glazed bores must be honed. Use a non-precision hone such as this bead hone to *break the glaze.*

Now, rather than the full face of the skirts taking the thrust load, thrust is taken by the considerably smaller area formed by the ridges. The result is a rapid wearing of the knurls, or ridges—you're soon back to square one.

Knurling is temporary. Use it only if you are patching an engine to last no more than 5000 miles.

Glaze-Breaking—If you are going to bore your block there is no need to do any *glaze-breaking,* or restoring the *cross-hatch* pattern to the burnished cylinder-wall surface. This will be done during boring and honing. If you are reusing the original bores and pistons, you need to break the glaze. Cross-hatching provides good oil retention on the cylinder walls, which is necessary for piston-ring sealing. It also makes for quick ring break-in.

Glaze-breaking is done with a hone, but not with just any hone. A *precision-type* hone must not be used for glaze-breaking. It would do more than break the glaze. A precision hone will work to remove the taper in the cylinder. The bores would end up way over their wear limit. Instead, use a *bead*—sometimes called a *ball* or *brush hone*—or *spring-loaded* hone that *follows* the existing cylinder contour. These hones will not remove large amounts of metal from the cylinder as would a precision hone.

Block Decking—Don't send your

Block is being *top-decked.* **A few thousandths milled off each deck ensures they are true to the crankshaft-bearing bores. Standard procedure on racing blocks, decking is unnecessary on 99% of all rebuilds.**

block to the machine shop until you check the deck surfaces. Your main concern is that the decks are truly flat and damage free. If there are no signs of head-gasket leakage, or there are no large scratches or gouges in the decks, they should be OK. Warping is usually confined to the cylinder heads.

If the engine did have a head-gasket leak, check the deck/s for flatness. This is done with a straightedge and feeler gages. The proper straightedge is a large steel bar that has been precision ground—*don't use a yardstick!* If you don't have such a straightedge, ask the machine shop to check the decks for you. They won't charge much, if anything, to do it.

The actual measuring job is simple. Lay the straightedge on edge over the length of the deck, then try to pass a 0.003-in. feeler gage under it. If the gage will go under the straightedge, there is a depression in the deck. If a 0.005-in. gage fits, the deck surface requires milling—or *decking.*

Double-check that the decks are free of gasket material or dirt before you make your final decision. Take measurements at several different locations. Place the bar horizontally and diagonally on the decks at different locations.

Flat decks are doubly important if your machine shop is using a Van Norman-type boring bar that locates directly on the deck. If a deck is not flat, the cylinder bores will be machined off-center—they won't be perpendicular to the crank. Also, the decks should be free of any irregularities that could affect the boring bar. Remove any bumps or nicks with a large flat file. Be careful not to nick or gouge the bores.

If one deck has to be milled, mill the other one. This will prevent the cylinders on one bank from having a higher compression ratio than those on the opposite bank.

Decking a block raises the compression ratio slightly. How much it is increased depends on how much is removed from the deck; the engine may require higher-octane gasoline as a result. I suspect that finding gasoline that wouldn't ping was already a problem, so don't deck the block just to clean it off. And don't remove more than absolutely necessary if decking is required.

Notching Damage—If your engine was run for quite a while with a blown head gasket, there's a chance that a notch was cut into one of the decks. Usually, a notched block cannot be repaired by simply decking it. It must be welded.

Cast-iron is difficult to weld and requires nickel welding rod. It must then be *ground* rather than machined because of the hardness of the weld.

Consequently, repairing a notched block is expensive. Aluminum blocks can be heliarced, then conventionally machined—not quite as expensive.

If your cast-iron block is badly notched, talk with your machinist. It may be less expensive to purchase a new or used block rather than making the repair. Check the price of big Chevys in the local junkyards; it may be cheaper to buy a complete engine just for its block!

Main-Bearing Bores—It's time to give your block main-bearing bores some attention. Occasionally the roundness and alignment of the main-bearing bores will change, but it is uncommon.

Check alignment first. Make sure you do this if the crank didn't spin freely when you checked it during teardown, page 59. A simple way to check bearing-bore alignment is to install a straight crankshaft. Make sure the crank is straight using the check, page 77. Then turn to page 113 for installing the crank—don't install the rear-main seal. Once in place, spin the crank by hand to make sure it turns freely. If not, the main-bearing bores are out of alignment. They must be align bored or honed.

A more accurate checking method is with a special mandrel that accompanies an align-boring machine. The problem is very few machine shops have one of these, so many machine shops won't be able to align-bore your block. Second, getting a block align bored correctly is difficult.

My advice is not to worry about the main-bearing bores unless your block shows obvious problems: a main-bearing cap was loose in its registers or the crankshaft didn't spin freely in its bearings. Because accurate align-boring is hard to come by, impossible in many towns large or small, another block is often the best solution.

Just in case you want to check the main-bearing bores, you'll need a 2—3-in. mike and a telescoping gage or dial bore gage. Install the main caps, less bearing inserts, in order and in the right direction. Oil the main-cap-bolt threads, install and torque them 95 ft-lb; 110 ft-lb if it has four-bolt main-bearing caps.

All main-bearing-housing bores should measure 2.937—2.9380 in. all the way around—they should be round. If a bore is not to spec, the bearing insert will be crushed too

Another rarely needed machining operation is align honing—shown—or boring. Neil at Dragmaster has found that very few big-blocks need the bearing bores remachined, but was happy to demonstrate the operation for the camera. Big-blocks that experienced extreme detonation may need this step, so don't discount it entirely.

ALIGN BORING & HONING

Align boring or align honing the main-bearing bores is a machining process done in the course of building a racing engine. This automatically ensures that the bearing bores are in perfect alignment. It also automatically adds to the cost of the engine.

Remachining the main-bearing bores is not a normal part of an engine rebuild, simply because it is rarely necessary. Align boring or honing is only required when the original main-bearing caps are not used, or when a bearing bore is out of spec or damaged.

much or not enough. Either way, the bores should be align bored or honed.

Ring Types—Before you truck your block off to the machine shop, some words about piston rings are in order. There are three basic piston-ring styles available for the big-block Chevy: plain cast-iron, chrome and moly-filled. Stay away from cast-iron rings; they have poor wear qualities.

Chrome rings are recommended for an engine operated in extremely dusty conditions, but they have disadvantages. Unless you are rebuild-

ing an engine for a truck that will run around on dirt roads, chrome rings are not for your engine. The big disadvantages are a long break-in time and an inability to conform to cylinder-wall variations. Both are a result of the hardness of chrome, which give it long life in dirty conditions but reduce ring flexibility.

Moly-filled rings are the right choice 99% of the time. That's why Chevy uses them as original equipment. They last much longer than plain cast-iron rings, and break-in is virtually immediate. Chrome rings require thousands of miles to break-in; sometimes they never do. The secret of moly rings is their surface makeup. Under a microscope, a moly-ring surface shows many small indentations, or voids. These fill with oil so the ring always carries sufficient lubrication. They serve the same function as cylinder-wall cross hatching, but do a better job of it.

Chrome rings are very smooth and carry little of their own lubricating/sealing oil. Most of this oil is on the cylinder walls, in the hone marks. When the piston moves down to BDC on the power stroke, the cylinder wall is exposed to the burning air/fuel mixture. Consequently, much of this oil is burned away. Therefore, when chrome rings move back to TDC they receive minimal lubrication.

Moly rings, on the other hand, carry their own oil supply. This oil film keeps the rings from dry contact with the cylinder wall, reducing piston-ring and bore wear. It also enables them to seal more effectively, thus increasing power and reducing blow-by.

After choosing the rings for your engine, the next decision is made for you: cylinder-wall finish. Moly-coated rings work well with a fine finish—even to the point of being polished—but chrome rings work best on a coarser, oil-retaining finish.

For moly rings, hone bores with a 400-grit stone; chrome rings should be fitted in bores finished with nothing smoother than a 280-grit stone. With either finish, the cross-hatch pattern should be the industry standard of 60°.

So, before you drop off your block to be machined, decide which rings you'll be using. Your machinist will finish the bores accordingly.

Modern boring bar locates off main bearings; others locate off block decks. The decks must be free of imperfections if the later-style boring bar is used. Block is being bored 0.060-in. oversize in three steps: 0.030, 0.025 and 0.003 in. This leaves 0.002-in. *honing stock* to ensure fractured material will be completely removed by honing. Inside mike is used to check bore diameter before final cut.

Install Main-Bearing Caps—The last thing to do before the block goes to the machinist is to install the main-bearing caps—if they aren't already in place. Install them in their proper order and torque the bolts 95 ft-lb; 110 ft-lb if it's a four-bolt block.

The installed main-bearing caps stress the block so the bores will be round after they are machined and the engine is assembled. If the block is honed without the caps in place, the bores will be out-of-round when the caps are installed. It is important that the correct amount of torque be applied to the main caps or the strain on the block will be incorrect.

To torque any bolt accurately you need a *torque wrench*. You can rent one if you don't own one. There are two basic styles: *beam* and *clicker*. The beam type is simple, less expensive and reliable. The clicker type is expensive, but easier to use.

There's another aspect of stressing the block for boring and honing. When you think about it, if it's necessary to install the main caps, what about the cylinder heads? Well, although they do make a difference, they don't make as much of a difference.

Bores are honed to size following boring. Main-bearing caps must be torqued in place for honing to produce truly round cylinders.

The first thing to do when you get your freshly bored block back from the machine shop is to inspect the top of the cylinders. Make sure each was chamfered (arrow). If not, chamfer them using a half-round or round file. Take care that you don't run the end of the file into a bore and scratch or gouge it.

After all machine work is completed, the block must be cleaned. Most shops will hot-tank and oil-coat blocks, but clean yours with a stiff-bristle brush, soap and water to make sure. Dry the bores with paper towels immediately afterwards and coat all machined surfaces with spray-on oil: cylinders, bearing bores, lifter bores and decks.

But how is a block bored or honed with the heads installed? They aren't; *block plates,* or *deck plates,* are installed instead. Nothing more than a thick slab of steel, a block plate has four large-diameter holes in it, each of which centers on the bores. These holes allow access for the boring bar or hone.

Head-bolts don't strain the block as greatly as the main caps, so using block plates is not necessary with a standard rebuild. And having your engine bored or honed with deck plates costs extra. So don't have it done, unless you have money to burn or your engine will be used for all-out racing.

Crack Check—One final thing you should do or have done is crack check the block. This can be done by *Magnafluxing* or *Spotchecking.* You'll have to have Magnafluxing done for you. You can Spotcheck a part at home.

Magnafluxing begins by magnetizing the part being checked. Iron dust is then sprinkled on the suspect area—such as around a core-plug hole. If a crack exists, the iron-dust particles are attracted to it. What may have been an invisible crack now shows up as a fine, gray line.

Spotcheck kits consist of a penetrant—a liquid red dye—and a developer—a white powder. Cleaner is also included to clean the surface prior to and after the check. After the area being checked is cleaned, it is sprayed with penetrant. After cleaning again, developer is sprayed onto the same area. If a crack exists, penetrant that has *penetrated* the crack will bleed through the white developer and show up as a bright red line.

Spotcheck can also be used to check rods, heads, pistons and crank. The major advantages of Spotcheck over Magnaflux are its ability to check non-magnetic parts, such as pistons, and its initial cost and convenience. Spotcheck is available from: Magnaflux Corporation, 7300 West Lawrence Avenue, Chicago, IL 60656.

Cracks mean the block may need replacing. Even if no machine work is required, have the block Magnafluxed.

Bore Chamfer—When you get your block back from the machine shop, inspect the tops of the bores. Each should be *chamfered*—angled by filing or grinding—so there is no sharp corner at the top of the bore where it meets the deck surface. Most machinists do this as part of the bore and hone job. If your block got through without being chamfered, do it yourself. Removing the 90° corner will provide a lead-in for the piston rings during installation. It will also eliminate a sharp corner that may overheat and cause preignition.

Chamfer the bores with a *rat-tail* or *half-round* file. Don't use a flat file or you'll be making lots of sharp edges. Use a light stroke to arrive at a 45° angle cut about 1/32-in. wide. Try to keep the file at an even 45° around the entire edge. And above all, *be careful.* Keep the file from contacting the opposite cylinder wall! You sure don't want to gouge a fresh cylinder bore.

Suds Time Again—You are probably tired of all this cleaning work—but the block has to be cleaned again after it comes back from the machine shop. It might look clean, and it should after being hot-tanked or jet-sprayed, but that was before the machine work. Run a finger up and down the bore and you'll feel the after-effects of machining—metallic and abrasive grit.

To prevent this grit from circulating through the lubrication system, you must clean the block. Otherwise grit will embed in the main, rod and cam bearings and get between cylinder walls and pistons, valves and guides and into all other critical, close-tolerance areas of your engine.

The final cleaning is a little different from the previous solvent sessions. This time soap-and-water is the cleaning agent. You'll also need a bucket with dishwashing detergent, stiff-bristle scrub brushes and a rifle-type brush.

Put the block in the driveway and have a garden hose close at hand run-

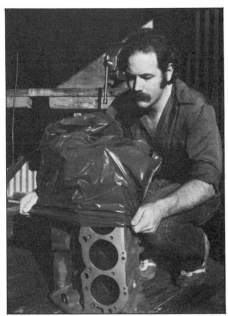

After oiling, cover block with a plastic trash bag to protect it from dust, dirt and moisture. After bagging, store block under the bench until you need it for engine assembly.

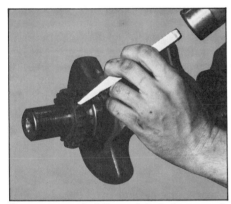

Remove sprocket if crank is to be ground. Don't use the hammer-and-chisel method illustrated. Rather, use a brass punch or puller to prevent damage to sprocket or number-1 bearing journal. Replace sprocket if it has high miles—30,000 or more—or if you're changing to a roller-type chain.

ning at a fast trickle. Hose down the block; then scrub every inch with the scrub brush and soapy water. Scrub the bores vigorously. A wipe through each bore with a white paper towel should come out perfectly clean. If dirt shows on the towel, keep scrubbing until the bore passes the towel test. Use the rifle brush in the oil galleries—followed by a shot of high-pressure water from the hose.

Drying with compressed air is best, but toweling works almost as well. Don't use cotton shop-rags or other cloth rags for drying. Cloth threads or lint could cause oiling problems later. Paper towels are better; paper frag-

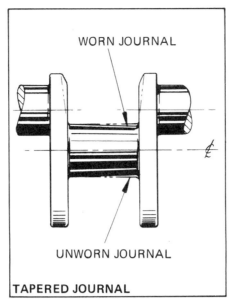

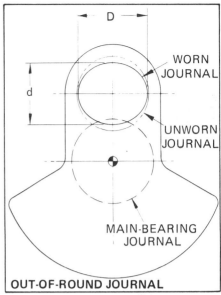

Check for taper and out-of-round when miking crankshaft. Out-of-round is the more common of the two, especially with rod journals. Drawing by Tom Monroe.

ments dissolve in engine oil with no ill effects.

As soon as you finish cleaning each bore, rinse it, dry it quickly and give it a shot of water-dispersing oil. This will keep that bore rust-free while you're scrubbing the others. Do the same to all machined sufaces. Redry and reoil the block after it is all scrubbed.

Whatever you do, don't walk away and let the block air-dry or you'll have rusty bores—and rust is abrasive. Truly clean metal—such as freshly honed bores—rusts very quickly.

When the block is cleaned and rust protected, that oily surface becomes a dust magnet. Turn the block on end and slip a plastic trash bag over it. Tie off the end to seal it. Store the block out of your way until assembly time.

CRANKSHAFT

Although a crankshaft is big and husky, it is a precision part. Big-block-Chevy cranks are no exception. Many are made from steel alloys usually reserved for heavy-duty race engines. Even ones not intended for racing show the careful layout and overall beef of a superior crank.

It would be a shame to treat such a critical engine part casually. Unfortunately, many mechanics don't check cranks carefully enough. This carefree attitude is the result of the remarkable service records chalked up by modern-day crankshafts. But problems do arise. So take time to make sure your crank is A-OK.

Most crank problems are caused by dirty oil. Although dirty oil is harder on the bearings than the crank, there is a limit to what a crank journal can take. Exceptionally dirty oil will scratch the journals. Deep grooving will result if large fragments are carried into the oiling system. Fortunately, damage of this kind can be repaired for comparatively little cost—considering the cost of a new crank. Only if the crank broke as a result of some really destructive internal problem, such as a thrown rod, should you have to replace it.

What Can Go Wrong? — Crankshaft-journal condition is most important to crankshaft/engine operation. As I just mentioned, the journals could be scored, scratched or grooved from dirty oil or foreign particles—or they could be worn undersize, *out-of-round,* or *tapered.* Additionally, the thrust faces could be worn, or there might be cracks hiding anywhere on the crank.

Start crankshaft inspection by checking the texture of the bearing journals. Let the bearing inserts be your first guide. If they show wear, then the crank will show at least a little. Run your finger over the journals. If there is any roughness you'll feel it. Run a fingernail the length of the journal—it will catch on any roughness.

Don't think you are doing something wrong if you can't feel anything. There's frequently no roughness to feel. This will be true if the engine was well cared for through regular and frequent oil and filter changes. Assuming

Checking out-of-round and taper. Make several measurements around each journal and along its length to make sure you get the right measurements. After making a measurement, record your findings; you'll never remember them if you don't.

journal sizes are OK and the crank is straight, all you need to do is polish the journals to ready the crankshaft for installation.

You may feel deep scratches, or scoring. This means the crank must be *ground*—the main- or rod-bearing journals ground to a smaller diameter. How much grinding depends on how bad the journals are. Most cranks clean-up with a 0.010-in. reduction in diameter, but 0.020- and 0.030-in. are common reductions.

After a crank has been ground, *undersize* bearings must be used. Undersize refers to journal diameter, not the bearing insert. The bearing must be thicker by the amount removed from the bearing journal to take up the extra *oil clearance*—distance between the journal and the bearing.

Ask your parts man for *undersize* bearing inserts to compensate exactly for metal removed from the crank—or 0.010-, 0.020-, 0.030- or 0.040-in.-undersize main or connecting-rod bearings. Double-check all measurements to avoid mismatching inserts and journals.

Even if only one journal is bad, *all* the main or rod journals must be ground. Example; the crank is OK except for one bad rod journal. All rod journals are ground the amount that is required to clean up the one bad journal. The reason for this is basic. Bearing inserts are sold in sets; all bearing inserts in that set are the same size. Therefore, it doesn't make sense to grind only one journal.

You won't save money on the machining job either. The cost of setting up the crank in the grinder represents most of the cost. And it is the same for one or four or five journals. You'd have trouble getting the machinist to do only one anyway. However, you can have the mains ground and not the rod journals, or vice-versa. This is standard practice. Or you might grind the rods 0.020-in. and the mains 0.010-in.

Out-of-Round—Bearing journals do not necessarily wear evenly, especially rod journals. They tend to wear into an elliptical or oval shape. It's easy to see why when you consider the load reversals applied to a rod-bearing journal. Load changes occur as the piston goes through its four strokes, or cycles.

The rod and piston push against the journal during the compression and exhaust strokes, and particularly the power stroke while at TDC. The rod and piston pull against the journal during the intake stroke. Inertia loads change with the angle of the crank pin, or rod journal, and the relative movement of the piston in its bore. Rather than go further with this discussion—it can get involved—I believe you get the point: Varying loads cause uneven bearing-journal wear.

To detect wear—even or uneven—you need a 2—3-in. outside micrometer. Measure the journal in several positions around it, but in the same plane. If your measurements vary, the journal is *out-of-round*. Ideal-

ly you should be able to make a measurement, then move the mike around the journal without changing its setting.

To determine how much a journal is out-of-round, subtract the *minor dimension*—smallest measurement—from the *major dimension*—largest measurement. The resulting figure must not exceed 0.001 in. If out-of-round exceeds 0.001 in., you must have the crank ground or replaced to correct the problem.

Main-bearing journals can also be out-of-round, so check them.

Taper—When a bearing journal wears front-to-rear, or along its length, it is said to be *tapered*. Measure the journal along its axis. If these measurements vary, the journal is tapered.

Excess taper must be corrected. If it isn't, the bearing insert will be loaded more at the big end of the journal than at its small end. In effect, the load that should be distributed evenly over the length of the bearing will now be concentrated in a much smaller area. The bearing design load will be exceeded and it will wear out quickly.

Taper is expressed in *thousandths-of-an-inch per inch*. So, make your first measurements at one end of the journal, then move one inch over and take another. Subtract the smaller from the larger to find the taper. Maximum-allowable taper is 0.001-in. per in. for both mains and rods.

Journal Diameter—Now that you've checked taper and out-of-round,

Polishing crank journals with 400-grit emery cloth. There isn't much more to the normal crankshaft reconditioning. To complete the reconditioning, oil passages should be rodded out and journals protected with spray-on oil.

To check runout, install crank with front and rear main bearings only. Set up dial indicator. If total-indicated runout exceeds 0.007 in., have crank reground or trade it on a crank kit.

check journal diameters. Front four mains should measure 2.7481—2.7495 in.; rear should be 2.7478—2.7488 in. Connecting-rod-journal standard is 2.1985—2.2000 in.

Be aware that the crank could be undersize if the engine has been rebuilt. This is probably the case if the main or rod journals are grossly undersize by the same amount and in increments of 0.010 in.: 0.010, 0.020 or 0.030 in.

Subtract journal measurement from the above standard diameters to determine journal undersize. Keep track of this for ordering bearings. If you find that the journals are not worn the same or don't fall within a given undersize, have the crank reground to the next undersize.

Crank Kits—If your crank needs regrinding, you can save some money. Buy a *crank kit* rather than have your crank custom ground. A crank kit consists of the crankshaft *and matching bearings*. Make sure the bearings are included; they represent your savings. You exchange your crank when buying a kit, so there's no wait—important if you are in a hurry.

The only disadvantage with buying a crank kit is if you have a heavy-duty, forged crank. Make extra sure the one you get in exchange from the rebuilder will also be forged. Otherwise have your crankshaft reconditioned by a

local machine shop. If you don't pay attention to the small print on your sales contract, you might end up with a cast-iron crank in trade for your forged one. Someone will make out on this one—it won't be you.

The crankshaft rebuilder is not intending to do anything dishonest because the crank he will give you in trade will bolt into your engine; and this is what he promised. If you're rebuilding an LS-6, have that forged, cross-drilled and Tuftrided crank worked over by a machine shop. Don't risk losing it.

Tuftriding—While on the subject of heavy-duty cranks, you need to know about Tuftriding. Many big-block Chevys came with forged cranks. Of these forged cranks, a few require an additional step in their reconditioning. Specifically, high-output big-blocks with forged, *Tuftrided* cranks have to be treated specially after machining.

Tuftriding is a chemical heat-treating process that applies a very thin hard surface to the bearing journals. You can recognize a Tuftrided crankshaft by its dull-gray finish. When the journals are ground, this hard surface is removed.

If you want the crank to be as good as new, it must be Tuftrided again. Be sure to relay this to your machinist or it won't get done. Just make sure you

know what the cost will be before you give him the OK; you may change your mind. However, if your crank is in good shape, all you have to do is polish the journals; there is no need to reTuftride.

Runout—Before declaring your crank OK or sending it off to have it reground, check its *runout*. Runout is simply how much a crankshaft is bent. You'll need a dial indicator and a supporting base to measure runout. With the block on its back, lay the crank in it with only the front and rear main bearings installed. Oil the bearings before you install the crank. Install the caps and snug the bolts. Set up the dial indicator next to and in contact with one of the center three main-bearing journals. The indicator plunger must be 90° to the journal and positioned so the journal oiling hole will miss the plunger tip as the crank is turned.

Turn the crank slowly by hand and watch the dial indicator for movement. When the indicator is at its *lowest* reading, zero the indicator. Turn the crank again until you find the *highest* reading. Read runout directly. Check runout at the other two center journals. Pay special attention to the center main—it should have the most runout.

Maximum permissible runout is 0.007 in. If your crank yields a runout greater than this, all is not lost. A crank may bow slightly when it is unbolted from the block. You may be able to correct this by bolting it back in. Rotate the crank so its point of maximum runout is away from the block. With only the cap-half of the center-bearing insert installed, torque the cap bolts. Leave the assembly undisturbed for a couple of days and remeasure.

If runout is now within limits, great. If not, replace the crank. Or if you prefer to keep the original crank, for one reason or another, search out a shop that uses a hammer technique to straighten cranks. This job is for experts only, not the local blacksmith. If you don't live in a large city, you'll probably have to ship your crank away for this service. Above all, don't use or let anyone use a hydraulic press to do the straightening job. This will straighten the crank, but it may also *crack* it.

Crack Checking—Now that we're on the subject of cracks, have your crank Magnafluxed or Spotchecked. This is especially important if it had to be straightened. If a crack is found, replace the crank.

Post-Regrind Checks—After the crank returns from the machine shop, give it a few checks. Begin by checking the oil holes. The edge of each oil hole must be chamfered so it blends smoothly into the journal surface. A sharp edge will cause trouble. It could cut a groove in the bearing. If the oil holes are not chamfered, return the crank and voice your complaint.

Another area where improper regrinding can cause trouble is at the journal fillet—where the journal blends into the counterweight or throw at each end. The Automotive Engine Rebuilders Association's Crankshaft Manual tells the story pretty well. It seems the auto companies provide detailed specifications on crank grinding, but the limits are fairly wide.

Grinding everything that comes through the door with 3/32-in. fillet radius satisfies almost all requirements. Exceptions are listed in the manual; the big-block Chevy is not one of them. So you stand a good chance of having your big-block-Chevy crank ground with a 3/32-in. radius. It will work fine.

Fillet radii serve a purpose; to reduce the possibility of cracks developing at the ends of the journals. If a radius is too small, a crack could start in the sharp corner. The crack eventually causes a break. In this respect, the bigger the fillet radius, the better. However, a fillet radius that is too big

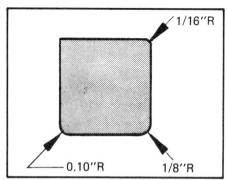

Use pattern for making a template to check bearing-journal radii. Journal radius should not exceed 0.100 in., otherwise it will edge-ride bearing.

Much quicker than the shoe-shine method, machine shops use an electric-powered polisher and lathe to polish crank journals.

will *edge-ride* the bearing. The flat surface of the journal would then be too narrow to accommodate the entire bearing.

You may have noticed that the old main-bearing inserts showed a thin line of babbit around their edges. This was the result of edge-riding, but not caused by an overly large fillet. On high-mileage engines, it is usually caused by excessive crank *float,* or *end play*—excessive front-to-rear crank movement. Although it could be caused by crank thrust-face wear at the number-5 journal, it's usually caused by thrust-bearing wear.

Edge-riding, like that found in worn engines, is more of a problem on standard-shift cars. This is because force required to release the clutch pushes the crankshaft forward against its thrust bearing with each shift.

Check crank-bearing-journal radii with a template made from the pattern on this page. An alternate, but cruder, method is to fit half of a new bearing insert to the appropriate journal and check that there is no interference between the journal and the fillet. If there is, the fillet is too large or small: Return the crank to the machine shop. Running it will only result in disaster.

Final Polish—If your crank doesn't need to be ground, reconditioning it is simple. Polish the journals and smooth any small scores or dings. A *needle file* works well to remove any projections caused by dings. While you have the small file in hand, check the oil holes. Smooth any sharp edges.

Even if you have a reground crank, polish its journals. Use narrow strips of 400-grit emery cloth as shown to do the polishing. You'll find emery cloth

at a hardware or auto-parts store. It comes in long narrow strips; 2-ft long by 1-in. wide is about right. If it's 2-in. wide, tear it down the middle.

Set your crank on the bench so it won't roll around, then loop the strip over a journal. Start at one point on the journal and move around its circumference, ending up where you started. Use a light, even pressure and take care not to stay in one spot too long. You don't want to "flat-spot" the journals! The idea is not to remove metal, but to clean and resurface the journals. Do the oil-seal area as well, but don't bring it up to a bright polish. Like the cylinder walls, the seal area should have some tooth to carry oil and lubricate the seal. This increases seal life and improves its sealing.

Final Cleaning—Now that crankshaft reconditioning is complete, make sure it is clean. Even if you had the crank hot-tanked, clean out the all-important oil passages. Make sure they are squeaky-clean. Pull a piece of rag soaked with solvent, lacquer thinner or carburetor cleaner through the oil passages with a wire.

Nylon coffee-pot brushes or small gun-bore brushes work great for cleaning crankshaft oil passages. If you used one to clean the block oil galleries, it will work fine. This is a necessary step, because even a hot-tank won't get all the crud off of or out of a part. Also, use a stiff-bristle brush and solvent on the counterweights if they look dirty. Finish the cleaning with paper towels.

Give the crankshaft the same rust-prevention treatment you gave the block. Spray its bearing journals with oil. Unless it is used with a manual

A gun bore brush run through each oil hole will remove any residual dirt or grit. A blast of compressed air should follow. To finish, wipe the crank clean with paper towels and protect it against rust with light oil.

Removing a pilot bearing with a slide hammer. Weight is bumped against slide-hammer handle to pull bearing from crank. Install the new one with punch that matches bearing's outer perimeter.

Sintered-bronze pilot bushing is self-lubricating.

transmission, slip a plastic trash bag over it and store the crank out of harm's way. If a manual transmission is used, you have one more chore.

Pilot Bearing—Crankshafts used with a manual transmission have a *pilot bearing,* or *bushing,* installed in the rear of the crankshaft. A pilot bearing is an oil-impregnated sintered-bronze sleeve, or *Oilite* bushing. This bearing supports the forward end of the transmission input shaft. If it's a replacement crank, it won't have one. You'll have to install a new bearing.

Crankshafts used with automatic transmissions won't have this bearing. If your car uses an automatic, just make sure there is no pilot bearing there. If there is, remove it. Read on for how to do this. If you install the engine with a pilot bearing left in the crank, severe automatic-transmission damage will result.

The bearing inside diameter (ID) is important. If the pilot-bearing ID is badly worn, the transmission input shaft will flop around. This may cause the clutch to chatter and the transmission to pop out of gear.

One quick measurement can be made by inserting an extra input shaft into the bearing and checking for side movement. Very slight movement is OK, but anything more is cause for replacement. A more accurate measurement can be made with a telescoping gage and 0—1-in. micrometer, but my advice is not to worry about measuring it. Replace it. Cost and work are minimal when you consider the problem of replacing the pilot bearing later.

The quick and easy way to remove a pilot bearing is with a *slide hammer*—a rod with an interchangeable pulling tool at one end, a handle, or *hilt,* at the other, and a sliding weight in between. The weight is impacted against the handle for pulling. Body shops frequently use these to pull out dents. A hook-type tool is used for pulling pilot bearings.

Insert the the hook into the bearing and catch its backside, slam the slide-weight into the hilt a couple of times, and the bearing will pop out of its bore.

Drive in pilot bushing until it bottoms. If you don't have an old input shaft, use large-diameter punch or socket to install bushing. Be careful that you don't damage the bushing bore.

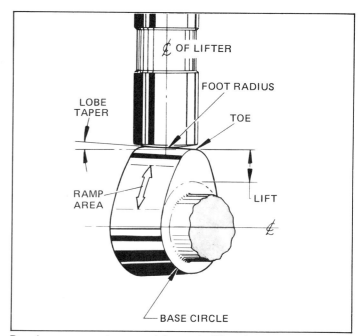

Drawing exaggerates cam-lobe and -lifter features. Lobe and lifter are designed so that lifter rotates in its bore and there is maximum contact between lobe and lifter. If any cam-lobe toe is worn across its entire width and the corresponding lifter foot is worn flat or concave, camshaft and *all lifters* should be replaced. Drawing by Tom Monroe.

Front section of old manual-transmission is useful for driving in new pilot bushing and for installing clutch.

A slide hammer is too expensive to purchase for this one-time use. So if you can't borrow or rent one, try the *hydraulic* method. Start by filling the bearing bore and the cavity behind the bearing with grease. It must be full of grease and purged of all air.

Fit a snug-fitting dowel into the bearing. Drive the dowel into the bearing with a heavy hammer. This will pressurize the grease and drive the bearing out. As the dowel goes in, the bearing will come out. Clean all but a light film of grease from the crank. This will help with driving in the new bearing.

To install the new bearing, you'll need a brass punch that is slightly smaller than the bearing outside diameter (OD). A deep socket will do if you're in a pinch and you don't mind hammering on an expensive tool. Use a brass mallet to limit

damage. Hold the bearing square against its bore in the crank, place the punch against the bearing and drive it in.

Start with light hammer blows and drive the bearing in straight. If it starts to cock, remove the bearing and start over. Don't keep hammering, hoping it will straighten out—it won't. You'll destroy the bearing. Once it is well started, drive it in until it bottoms.

CAMSHAFT & LIFTERS

After all that bottom-end work, the cam and its lifters should provide a change of pace. Unlike the "bulletproof" crank, a cam and its lifters are more than likely worn out.

It is important to follow the inspection procedures I've laid out for you because of the fast-wearing and critical nature of a cam and its lifters. Slipping up now could mean trouble the moment you fire up your new engine, so don't ignore steps.

First, memorize these absolute rules for servicing cams and lifters:
1. When reusing a cam and its lifters, never install the lifters out of order. Lifters must be reinstalled in the same bores, *in the same engine* and on the same cam lobes. If you are not sure which bores any of the lifters came out of, replace all of the old lifters.

2. Never install *old* lifters with a *new* camshaft.
Cam & Lifter Design—Both the above rules are a result of the high working pressures between lifters and camshaft lobes. To understand completely why it is so important to follow these rules, a short course in cam and lifter design is in order.

Let's start with some definitions. The high point of a cam lobe is the *toe*. The opposite side, centered 180° from the toe, is the *heel*. This section is also called the *base circle* because it is the circle that the lobe is built on. The shape of a cam lobe when viewed from the end of the cam is called its *profile*.

Cam profile is designed by engineers using complex mathematical formulas. Many more subtle aspects of a cam profile are not easily detectable by the eye. One of these near-hidden features is called the *rake angle*. Rake angle is the angle the surface of the lobe makes with the axis of the camshaft. It is approximately 1°.

Lifter bottoms, or *feet,* are ground on a 30-in. *spherical* radius. This slightly convex shape is hard to see, but if you butt the bottoms of two lifters together, you should be able to rock them slightly. If they don't rock, the foot radius of one or both is worn off.

When you consider lobe-rake angle and lifter-foot radius, you'll see that the lobe and lifter contact area is smaller than their total area indicates. However, because of production tolerances, this ensures a *minimum* contact area between lobe and lifter as opposed to a lobe with no rake angle and a flat-foot lifter. Additionally, the lobe contacts the lifter slightly off-center to rotate it. This keeps the lifter from wearing in one area. Examine the drawing for a clearer understanding of the cam-and-lifter relationship.

CAM & LIFTER INSPECTION

As with any other engine part, you are trying to estimate the amount of *unused* miles left on the cam and lifters by inspecting them. You measure how much wear has taken place in the miles already accumulated to determine wear rate, then subtract this from the expected mileage. However, when it comes to the valve train, determining what to use or scrap is pretty simple; buy a new camshaft and lifters to get the most mileage possible with the least chance of failure.

Unlike the bullet-proof bottom-end of the big-block Chevy, the camshaft and lifters wear fairly rapidly. Reusing these components, especially the lifters, can be risky. It is common for a cam or lifters go bad in just a few minutes running on a freshly rebuilt engine. As a final word of warning, if you lost the order of your lifters buy new ones! It is absolutely imperative that the same lifters mate-up with their original cam lobes. If you don't know the correct order of *all 16* lifters, your chances of installing them in the correct order is 20,922,789,890,000 to 1! Not good.

If you intend to reuse the cam and lifters, be aware that you may not save much in the long run. The lifters and cam will continue to wear, but at a faster rate than if both were new. So don't be surprised if you have to perform a cam-and-lifter change within 20,000 miles. You've come this far with the rebuild, so replace the cam and lifters.

Camshaft First—If you insist, let's see if the cam and lifter are reusable. Every cam/lifter inspection starts with the cam. If it is bad, you must change both cam and lifters no matter how good the condition of the lifters. Remember the second rule; *never in-*

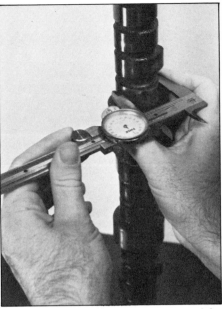 

To find cam-lobe lift, measure across lobe base circle, then turn lobe or caliper 90° and measure across the toe. Subtract to find lobe lift. This method is not accurate with high-performance camshafts because the opening and closing ramps extend into the base circle. The apparent lift will be less-than-actual. This method, however, is accurate when comparing lobes. See chart on page 13 for lobe and valve lift.

When inspecting a cam lobe, check for wear extending across entire lobe, especially the toe. These lobes are shot as evidenced. Compare worn areas to unworn areas (arrows).

stall used lifters on a new cam.

If you checked valve lift with a dial indicator during the diagnostic session, there is no need to check *lobe lift* now. You can refer back to your notes and see which lobes were worn, if any. Remember, to be accurate you should've measured with a dial indicator on the pushrod. If you didn't measure the cam lobes before, do it now. You'll need a pair of vernier calipers or a micrometer to measure lobe lift.

Measuring the lobes is a comparison check, one lobe to another. Unlike the check in the diagnosis chapter, the purpose here is is to compare relative lobe wear, not lift. Because cam lobes

wear at a rapid rate *once they start wearing,* you should have little trouble pinpointing a bad lobe.

Take the first measurement across the lobe, halfway between its heel and toe; this is the *minor dimension*. Then measure the major dimension, from the tip of the toe to the heel as shown. Both measurements should be made at the points of maximum visible wear and the same distance from the same edge of the lobe. Now subtract the minor dimension from the major dimension to get *approximate* lobe lift.

You can't depend on these measurements to reflect lobe lift accurately, particularly with a high-performance cam. This is primarily because you'll be measuring across the ends of the *opening* and *closing* ramps—transition sections of the lobe between the toe and the base circle. The reading you get will be greater than base-circle diameter. As a result, lobe lift will appear to be less than actual.

Measure all 16 lobes, recording your results as you go. To compare your findings, place all lobe-lift figures for the intake lobes in one column, and all the exhaust figures in the other. Intakes should be within 0.005 in. of each other; the same applies to exhausts. If there is *one* bad lobe, the cam *and lifters* must be replaced.

As for indicating which lobes are which, start at either end of the cam, the order is then:
EIIEEIIEEIIEEIIE.

Camshaft-Bearing Journals—Camshafts are more durable than cranks in one respect; their journals seldom wear out. Don't take this as an excuse not to check them. All should measure 1.9482—1.9497-in.

As with the crank, there is an out-of-round limit for the camshaft. However, if it turned easily in the old bearings, it'll be OK in new bearings.

To check the cam out of the block, support it with two V-blocks, one under each end journal. Or support it between centers, such as in a lathe. Indicate the center bearing journals as you did with the crankshaft. Runout should not exceed 0.0015 in.

The final check will be to install it in the block and spin it in well-oiled bearings. Unless the cam shows play or binds as you rotate it, there is no problem.

Lifter Inspection—You can skip this step if you're buying a new cam, because then *you must install new lifters.* At the risk of boring you, I repeat, keep those lifters in order or you must buy new ones! It is all too easy to get "lost in the woods" if you start moving lifters around at random, so go about your inspection in an orderly fashion. Do them one at a time.

Inspect the lifter foot for damage. Look especially for pits or depressions. If you find any, the lifter is junk. Don't bother going any further, just replace it. Chances are the others are in a similar state. Go ahead and check them too, but you'll probably be better off throwing them out and buying a new set.

GEAR-DRIVE CAMSHAFT

If you have a truck-type gear-drive cam to replace, you'll have to remove its gear and thrust plate, and install them on the new cam. An arbor press is needed, so it probably means a trip to the machine shop.

To remove the gear, support the cam vertically, cam gear up, in the press. Use a sleeve slipped over the cam lobes and butted against the backside of the gear hub to support the cam. Positioned directly under the press ram, push the cam out of the gear. One hand must be on the cam to catch it as it falls free of the gear.

Installing a new gear requires the same arbor press. Set a new thrust plate over the cam nose, followed by the gear. Line the gear up with the key and press the cam gear on until there is 0.001—0.005-in. clearance between it and the thrust plate. Check that the thrust plate rotates on the cam.

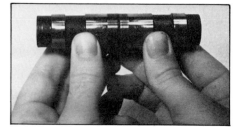
Hold two lifters foot-to-foot and try rocking them. Although slight, you should be able to feel a rocking motion if there is any radius left.

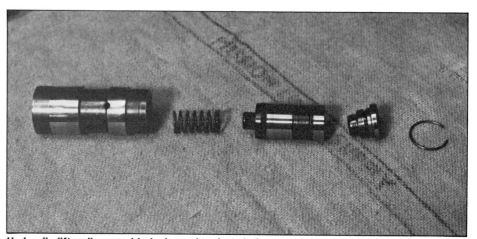
Hydraulic lifter disassembled, cleaned and ready for reassembly. If you have unlimited time and want to risk disassembling the lifters to clean them, do one at a time so you can't mix the parts. Otherwise, immerse each lifter foot down in a pan of solvent and pump the plunger up and down with a pushrod.

If a lifter is worn bad enough to be pitted, you should also see heavy wear marks. Look at its foot from different angles and watch the reflections. Make this check even if the foot appears to be OK. If you see a circular pattern at the center that reflects differently than the rest of the foot, that's an indication the foot may not be convex.

Double-check the foot with the "rock test." Butt one lifter foot-to-foot with another. If you can't rock them, junk them. A lifter worn flat or concave like this also has a worn cam lobe. Recheck the companion lobe toe for wear across its entire width. If this is the case, both cam and lifters are junk. However, if you get the rocking motion, you're in luck. Keep on checking.

Clean Lifters—If you are going to re-place the hydraulic lifters or your engine has solid lifters, there's no need to read this section other than to see how much "fun" you'll miss. However, if you are reusing the lifters, you do need to clean them.

There are several methods of cleaning lifters; the one you choose depends on how dirty they are. If your engine ran clean, there isn't much to do. Start by soaking them overnight in carburetor cleaner. This will clean the varnish off the outer surfaces, but won't do a thing for the internals.

To get at the inner parts, squirt solvent through the bleed hole in the lifter side. Lacquer thinner, carb cleaner or plain cleaning solvent works. As you squirt the solvent through the hole, work the lifter plunger with a free finger. This will get most of the goop out, but you can be sure there is a little residue left inside.

Another way to clean a lifter without taking it apart is to soak it overnight as suggested. Stand the lifter on a rag in a pan of solvent, then work its plunger up and down with a pushrod until it loosens.

If the plunger sticks when you push it down, hold the lifter upside down and give it a sharp rap against a piece of wood. If that won't budge it, soak the lifter in solvent and try again. If this doesn't work, use the final cleaning process.

Take the lifter apart. This involves a lot more time than just squirting solvent through its side or working the plunger up and down, but it is really effective—and time consuming. You must be extra careful about keeping the parts from each lifter separated. The most effective way is to *clean one lifter at a time.*

There are nine individual parts in a hydraulic lifter. To disassemble a lifter, remove the retaining ring inside the lifter shell with a pair of needle-nose pliers. Push down on the plunger with a pushrod to unload the retaining ring.

Next, pull out the plunger. If the plunger doesn't come out of the lifter body, work it up and down with the pushrod. If that doesn't work, pressurize it with the squirt can. Work the plunger up and down while squirting into the bleed hole; the plunger should eventually work free. Keep all the parts in order and clean each thoroughly.

Reassemble the lifter. Start with the internal pieces. The easiest way to reassemble the lifter is upside down. Make a stack, in order, of the internal parts. Set the lifter body over this stack after oiling its bore. Turn the assembly right-side up, use the pushrod to force the plunger down and install the retaining ring. After the first lifter is reassembled, go to the second.

Do all the lifters one at a time like this until all 16 are done. You might decide that all this work is more than you bargained for. If so, buy new lifters.

PISTONS & CONNECTING RODS

Piston Replacement—If you bored or are having your block bored, toss the old pistons away. However, if you can get by with honing the bores, you'll need to make sure the pistons are reusable.

Like the timing chain, sprockets, camshaft and lifters, pistons get a lot of abuse and wear; but they don't wear as badly. Although a well-cared-for, high-mileage big-block may have reusable pistons, most mechanics opt for "new" bores that can only be gained by reboring.

They reason correctly that fresh bores and new matching pistons will give them as-good-as or better-than-original engine longevity and performance. After all, used pistons and bores are just that—*used.* Some of their working life has already been spent. That brings us to the last part of the piston/bore relationship.

You may have decided the cylinders don't need reboring. But you may also find, after following the piston-checking procedures in this section, that the pistons should be replaced. If this proves to be the case, the reasonable thing to do is rebore and install oversize pistons.

New pistons are the greatest expense, so you won't have to spend much more to get straight bores. You'll also get a slight displacement increase. This will amount to about a 2-CID increase for every 0.010-in. increase in bore size. For example, a 366 bored 0.030-in. over will displace 5.60 more cubic inches; a 0.030-over 454 increases 6.43 cu in.; about a 1.5% increase in displacement for both. This is slight when you consider their overall displacement. What is important is you'll get the longest possible use from your rebuild with fresh cylinder walls.

INSPECTING PISTONS

If you are not going to rebore, inspect the pistons before removing them from the rods. If the pistons are OK, you can rebuild the engine without assembling and disassembling the pistons and rods. It is one less job, and will speed the rebuild. It may also mean one less trip to the machine shop to have the piston pins pressed out. One less bill too, and that's what this book is all about—reducing those bills!

Remove the old rings. There's an easy and a hard way to remove piston rings: by hand is tiring—and increases the chances of scratching pistons with the end of the rings. Using a *ring expander* is the best method. This inexpensive plier-like tool grips the ring ends and expands, or spreads

them apart, easily. Then you just lift the ring over the piston. There is less chance of damaging a piston by removing the rings with an expander instead of using your thumbs to spread them.

Both methods are easier when the rod-and-piston assembly is held steady by something other than your own hands. A vise is ideal, but be careful not to mark or bend the rod. To prevent this, clamp the rod between two pieces of wood. Soft lead, brass or aluminum vise jaws also work.

Clamp on the edges of the rod beam and position the piston so the bottom edges of its skirt rest on the vise jaws or wood blocks. This will keep the piston from flopping back and forth.

If you don't have a vise, clamp the rod and piston to a bench with a C-clamp. A C-clamp can also damage a connecting rod, so use a wood block between the clamp and the rod. If the bench is metal, put another block under the rod. Clamp the rod to the edge of the bench with the piston against it to steady the piston.

Piston Damage—Check for the following conditions during piston evaluation: general damage to the dome, skirt or ring-lands, or excessive ring-groove, piston-skirt or pin-bore wear. If any of these conditions exist, the piston must be replaced.

General Damage—Obvious damage is done to the piston when a valve drops into the cylinder, or when a piston-pin retaining ring breaks loose. You'll have to look harder for other types of damage. *Scuffing*, scoring or collapsing of the skirt, ring-land damage and dome burning can be found through a thorough visual inspection.

Scuffing refers to abrasion damage to the piston skirt. Scoring is like scuffing, but is limited to deep grooves. Both types of damage are caused by insufficient lubrication, very high operating temperatures—causing skirt-to-bore contact—or a bent connecting rod.

The common denominator in all three conditions is excessive pressure and temperature at the piston and cylinder wall. The most common cause of scuffed pistons is overheating. A stuck thermostat, blocked water passage, broken fan belt or operating the engine without

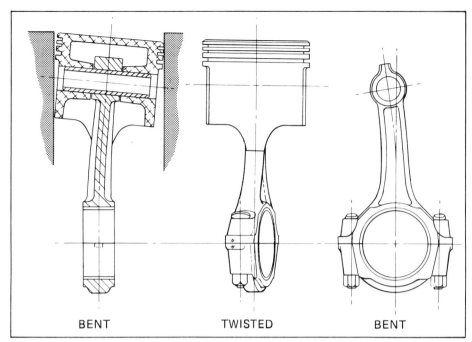

Aligning a rod. A vernier scale shows whether the rod is twisted or bent and how much. A large pry bar and notches at base of fixture are used to straighten rod.

BENT TWISTED BENT

Uneven piston wear is one symptom of a bent or twisted rod. First rod wears piston most at TDC and BDC. Twisted rod wears piston most at midstroke. The third rod, however, has no effect on piston wear and can be reused if the bend is not visible to the naked eye. Drawing by Tom Monroe.

Measure piston skirts in same plane as pin, but 90° to it. Then measure skirts at the very bottom and subtract. Measurement at bottom of skirt of a cast piston should be about 0.0005 in. more. Forged pistons should measure no less than 0.0007 in. more across bottom of skirt.

coolant will cause piston scuffing. In any case, replace a scuffed piston.

In addition to visible scuff damage to a skirt, the shape and resulting piston-to-bore clearance may have been adversely affected. Additionally, a high-mileage piston will have lost some of its controlled-expansion qualities. And the loads that caused the scuffing also may have bent or broken—*collapsed*—the skirt. Ashcan it and start over.

Examine the bent and twisted connecting rods in the nearby drawing. If a rod is bent or twisted, it will cause the piston skirt to wear or scuff unevenly from side to side. A little variation is normal, but when a piston skirt is scuffed left of center on one side and right of center on the other, you can be sure the rod is bent or twisted. Making sure the rods are not bent is a normal part of machine-shop routine, but it never hurts to advise them of your findings. You can't check the rods accurately for straightness because you don't have the equipment—but you may have the evidence that they need checking.

Piston-Skirt Diameter—Measuring piston-skirt diameter is not so much to determine wear as it is to check for collapsed skirts. Before measuring and judging the results, you should fully understand what piston-skirt collapse means.

Cleaning piston domes goes a little faster if you start with an old compression ring. Get the tough spots with a dull screwdriver. Finish up with a wire brush, taking care not to touch the ring lands. If you have a bench grinder with a wire wheel, use it.

All big-block-Chevy—and virtually all—pistons are widest at their skirts. This wide-skirt design allows the piston to be fitted tightly in its bore for better oil control and piston stability. A stable piston runs quietly.

In normal service, piston and cylinder walls warm to their operating temperatures and the piston skirts expand to a given size and shape. But, if the engine overheats, the piston skirts continue to expand until piston-to-bore clearance is gone; they are forced against the cylinder wall and begin scuffing. The abrasive scuffing

action increases the piston heat even more, but the cylinder wall doesn't permit further expansion. The skirts are then overstressed, or overloaded, and change shape to conform to the cylinder contour. At the extreme, they break.

After the engine cools, the piston skirts contract smaller than before because they have been permanently deformed—they are said to have *collapsed*. The result will be greater-than-acceptable bore clearance, both during warmup and while at normal operating temperature. The piston will slap during warmup and, possibly, after the engine is up to operating temperature.

If you were to reuse pistons with collapsed skirts they would be noisy.

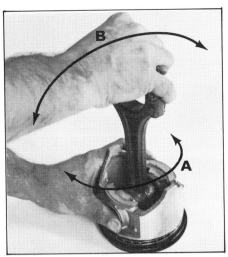

Checking pin-bore wear. First, soak piston in solvent to remove oil from pin and pin bore. Twist connecting rod in direction A, then try rotating it in direction B, 90° to its normal direction of rotation. If you feel movement, wear is excessive and piston should be replaced. Forged pistons are particularly bad in this area, so give them a thorough check. Photo by Tom Monroe.

If any have scuffed and/or collapsed skirts such as this one, rebore. Remember, the expense is not in the machine work, it's in buying new pistons.

If you don't have a ring-groove cleaner, break an old ring in half and grind one end as shown. Use it to clean the grooves.

The oil-ring grooves may have been deformed as well, providing poor oil control.

Heavy scuffing is a common sign of collapsed piston skirts, but to be sure, measure the piston in two places and compare the readings. First, measure across the skirt below the oil-ring groove and 90° to the piston pin. Then, move to the bottom of the skirt and measure again. Cast pistons should be 0.0005-in. wider at the bottom; forged pistons, 0.0007-in. If they are the same or narrower, chances are one or both skirts have collapsed. The piston is junk.

Inspect Domes—Because the melting point of aluminum is lower than cast-iron, the piston dome is first to suffer from heat caused by preignition or too lean a fuel charge. Such damage picks aluminum off the dome and deposits it on the combustion chamber, valves and sparkplug or blows it out the exhaust. This leaves the dome with a spongy or porous look that carbon adheres to.

Clean the piston domes so you can inspect them. If the carbon buildup is heavy, remove it with a dull screwdriver or old compression ring. Make sure the screwdriver is old and worn out. One with a sharp blade will scratch the piston easily. Obviously, you don't want to use a sharp-edge scraping tool on the pistons. Chisels and gasket scrapers are for chiseling

and gasket scraping, not cleaning pistons.

Follow the screwdriver cleaning with a wire brush, or wheel if you have one. Be extra careful to keep the wire bristles off the piston sides, particularly the ring grooves. Light scratches on top of a piston won't hurt, but they will cause trouble in the grooves and on the skirts.

Piston-Pin-Bore Wear—Two methods of pin retaining are used for big-blocks: pressed-in and floating with retainers. All but the highest-performance engines use pressed-in pins. You can have either cast or forged pistons with pressed-in pins.

Checking pin-bore wear differs with how the pin is retained. To measure a pressed-in pin and its bore, you have to press the pin out of the small-end of the connecting rod. Fortunately, there is a reasonably accurate and easy way to check for excessive pin and pin-bore wear without doing this.

To start, soak each piston in solvent to remove all oil. Work the rod back-and-forth while the pin is submerged in solvent. This will remove the oil cushion from between the pin and its bore. The piston-and-rod assembly should be at room temperature to get the most accurate check.

Hold the piston-and-rod assembly upside down on a bench, grasping the piston with one hand and the big end of the rod with the other. Hold the

piston with the end of your thumb on one end of the pin and a finger on the other. Now try to rock and twist the rod without moving the piston as shown in the photo. Rock the rod in the direction of the pin centerline. Next, twist it in a circle concentric with the piston circumference. If wear is excessive you'll be able to feel pin movement at your finger tips.

Again, if you feel *any* movement, the piston-pin bore is worn. Chances are the remaining piston-pin bores will be worn too if the first piston you checked is excessively worn. If this is case, the answer is simple: Buy new pistons and rebore. But because of the cost, check the remaining pistons just to be sure.

Pistons and connecting rods with floating pins are easy to disassemble, so take them apart and measure the pin, piston and rod with a telescoping gage and micrometer. Use a pair of retaining-ring pliers to remove the retaining rings at each end of the piston pin. Slide the pin from the piston and rod. Next, mike the pin at each end where it fits into the piston. Record your measurements. Use a telescoping gage to measure piston pin-bore ID.

Subtract the pin diameter from bore ID to get pin-to-piston clearance. Floating pins should have a 0.0004—0.0008-in. clearance between pin and piston. If this clearance ex-

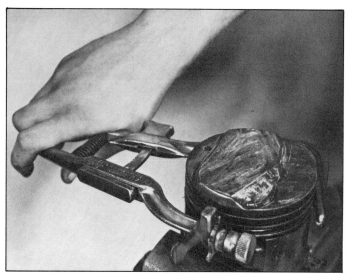

Cleaning ring grooves is much faster and more thorough with a ring-groove cleaner. If you do an occasional engine rebuild, this tool is worth the price.

Checking ring-groove wear with a ring and feeler gages. Consider 0.003-in. side clearance as maximum; 0.0025 in. is desirable. To check for consistent ring-groove width, slide ring and feeler gage around entire ring groove. If they bind or loosen, ring-land is bent or groove is worn unevenly; replace piston.

ceeds 0.0008 in., replace the piston. Measure the rod and pin in the same way. Pin-to-rod clearance should be 0.0005—0.0007 in.

In case you've taken apart your pressed-in-pin piston-and-rod assemblies, measure them. Piston-to-pin clearance should be 0.0003—0.0005 in., but this clearance can safely go up to 0.001 in.—no more. At the connecting rod, there should be a 0.0008—0.0012-in. *interference fit* between the pin and rod. The larger interference is more desirable, but should be not be exceeded because the pin will gall in the rod as it is pressed in. The rough, galled pin surface will then rapidly wear the piston-pin bore.

A word to the wise about forged pistons. Racers have found the pin bore to be a weak spot. If you are planning on reusing forged pistons, carefully inspect their pin bores and bosses. And if you are planning on racing your big-block, have these areas Spotchecked or *Zygloed*.

Zygloing is also used to crack-check nonferrous-metal parts. Similar to Spotchecking, the first step in Zygoling is to apply a liquid penetrant to the part. After penetrating for a few minutes, the surface is washed clean. Here's where the difference comes in: The part is then viewed under a black light. If cracks exist, they glow like a neon light.

Ring-Groove Cleaning—There's no way to measure ring-land wear accurately if the grooves are dirty. Two ring-groove cleaning tools can be used; one's free and the other's cheap. The free method uses the time-honored broken ring. The cheap method used a tool specially designed to clean ring grooves—a *ring-groove cleaner!*

If you are on a tight budget and don't want to spend the extra bucks for a "once-only" tool, use an old compression ring. Break the ring in half and grind or file one end as shown. Round all the ragged edges except the one cutting edge. This will keep you from removing metal from the ring lands. These surfaces must remain as tight and as damage free as possible for proper ring sealing.

The trouble with using a broken ring is that the ring will scratch and gouge the ring grooves if great care isn't taken. Once a ring groove is damaged, the ring cannot seal well between the ring and piston.

Because it is faster and there is less chance of damaging the piston, I prefer to clean ring grooves with a ring-groove cleaner. Cost of this tool is $10—30, depending on quality. As pictured, a ring-groove cleaner pilots in the groove as it is rotated. It does a spick-and-span cleaning job with much less danger of damaging the grooves.

Measuring Ring-Grooves—The easy way of measuring ring-groove width or ring side clearance is with a new ring and feeler gages. Problem: You don't want to lay out the money for new rings just to find that your old pistons are junk. Assuming you'd do the logical thing and rebore and install new, oversize pistons and rings, you'd be stuck with a set of rings that you had no use for. Therefore, use an old ring and compensate for its wear.

Original ring width for most production big-blocks is 5/64 in. (0.0781 in.) The L-88, ZL-1 and LS-7 engines use 1/16-in.- (0.0625-in.) wide compression rings. Truck big blocks use the standard 5/64-in.-wide compression rings in the top two grooves. The third compression ring is wider at 3/32 in. (0.0937 in.).

To measure ring-groove width, mike the thickness of one of the old compression rings. Subtract this from the *original* thickness measurement of that ring. Add this difference to the ring *side clearance* specified by Chevy: This will be the *checking clearance.*

Now insert the edge of the ring in the groove and gage the distance between it and the groove with feeler gages. You have the checking clearance right away, but you must also check around the groove.

Slide the ring and feeler gage around the piston and you can easily tell whether the ring lands are bent. The feeler gage will tighten-up in any constricted area and loosen where the ring lands are worn or bent away. You've just killed two birds with one stone, ring side clearance and ring-land condition.

When you find the gage that fits snugly—not tight or loose—subtract this measurement from the checking clearance. The difference is *piston-ring side clearance*—the dimension used to determine if the piston-ring grooves are serviceable.

Side clearance for all big-block compression rings is 0.0017—0.0032 in. Practically speaking, use 0.003 in. as the maximum-acceptable side clearance. You can't accurately measure that extra 0.0002 in. with a feeler gage.

Example: Let's say you have 5/64-in. (0.0781 in.) compression rings. You mike one of them and get 0.0771 in. Subtracting 0.0771 in. from the 0.0781 in. new-ring width leaves 0.0010 in.—about normal ring wear. Adding 0.001 in. to maximum-allowable side clearance of 0.003 in. yields a 0.004-in. checking clearance.

After measuring the ring groove with the ring and feeler gages, you find a 0.003-in. feeler gage fits best. The ring groove is OK because the checking clearance is under 0.004 in.

Use the same ring to check all pistons and grooves. Start with the top groove because it should be the worst, but don't skip the other compression-ring grooves.

Oil-Ring Grooves—Because oil rings and their grooves are so heavily oiled—compared to compression rings—they don't have a chance to wear out. Therefore, if the compression-ring grooves are OK, you can visually inspect the oil-ring groove in each piston for obvious damage and let them go at that. However, if you want to be doubly sure, measure the oil-ring grooves.

Oil-ring grooves are considerably wider than those for compression rings. To measure their width, stack two compression rings together and use them in combination with your feeler gages.

To establish an oil-ring-groove checking clearance, start by miking the combined thickness of the compression rings. Subtract this figure from oil-ring width of 3/16 in. (0.188 in.) and the maximum-allowable side clearance of 0.006 in., or 0.194 in. The difference is maximum allowable *feeler-gage* thickness, or *checking clearance.*

Example: If you stacked two 5/64-in. compression rings together and mike them, you should get about

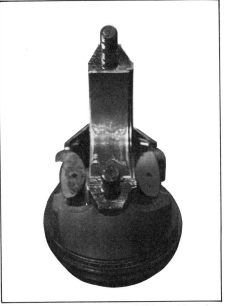
High wear at top center of this rod bearing is an indication of high combustion-chamber pressures caused by detonation.

0.156 in. Subtracting this figure from 0.194 in. gives 0.038 in.—the thickest feeler gage you should be able to get between the two compression rings and an oil groove. If you encounter the unlikely situation that an oil-ring groove is too wide, junk the piston.

Groove Inserts—You may have heard about this method of correcting worn ring lands. Let's say your engine is in pretty good shape, the cylinders don't need reboring, and the pistons are OK except for worn ring lands. Where do you go from here?

The pistons are chucked into a lathe and the ring grooves are machined wider. To compensate for the additional width, a *groove insert* is installed with the standard ring.

As with most low-buck fixes, it lasts about as long as it costs. Ring-groove inserting will cost you a little more than half of what it will cost to rebore and install new pistons, but lasts about half as long. If you don't need much service life from the rebuild and you want to save money "at all costs," this might be the way to go. But if you want to get the most for your money and want your big-block to last, do it the right way—rebore and buy new pistons.

INSPECT CONNECTING RODS

Doing an accurate job of inspecting connecting rods requires a special dial indicator and fixture. There's no way you can justify purchasing such expen-

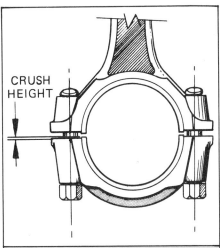
Precision bearing inserts are slightly bigger round than the bores they are installed in. Called *crush height*, about 0.001 in. of the bearing-insert-half ends project above the parting line. When the bolts or nuts are torqued, the bearing is *crushed*, locking it in place.

BEARING CRUSH

Precision bearing inserts have a slightly larger OD than the ID of bores they are installed in. When fitted, the ends of either insert half project slightly above the cap parting line. This distance, called *crush height*, is typically around 0.001 in. when the opposite end is flush with the parting line. When cap bolts or nuts are torqued, the insert halves are *crushed* to conform to the bearing bore. If the bore is too large, the bearing insert will not have sufficient crush.

Insufficient crush and the resulting looseness allows the bearing to move in its bore. And, if the insert becomes too loose, it can *spin* in the bore. This frequently destroys the bearing bore—the whole connecting rod in the case of a rod bearing—and crankshaft bearing journal in only seconds.

Reconditioning a rod restores its bearing bore size and shape. It also restores the honed surface that must "bite" into the back of the insert to hold it in place. To explode a myth, the *locating tang* or *lug* at one end of each bearing insert is not meant to keep the bearing from spinning. It's there for *locating*, or *positioning*, the bearing, as its name implies.

Make a quick check to see if a bearing was trying to spin by examining its backside. Shiny spots indicate the bearing was moving in its bore. This is a warning signal that the connecting rod probably needs reconditioning. It definitely needs its big end checked.

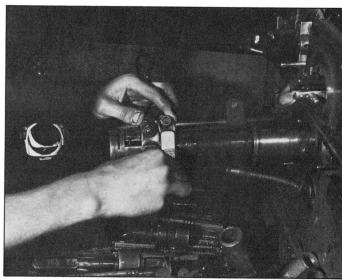

Torquing rod-bolt nuts prior to checking and, possibly, reconditioning big end. A rod vise is being used to hold rod assembly while it is being torqued. Assuming you don't own such a vise, clamp rod in a bench vise between two wooden blocks. Torque nuts on 3/8-in. bolts 50 ft-lb; 67 ft-lb for 7/16-in. bolts.

Resizing, or *reconditioning* rods by honing to restore bearing bores to correct shape and diameter. First, rod big end is measured with a special dial-indicator to determine whether resizing is needed.

sive equipment, so farm this job out to your engine machinist.

You can read the bearings and pistons, however, and determine whether a detailed rod inspection is absolutely necessary. For instance, if bearing and piston wear is even and your engine has been well cared for and not overrevved, you can *probably* skip checking the rods. However, because connecting rods are the most critical reciprocating parts in an engine, I suggest that you have them checked by a machinist, regardless of bearing- and piston-wear patterns.

Check Bearings & Pistons—The connecting-rod bearings should be in the rods just as you left them. Look square into the bearing bore. If the bearing inserts show excessive wear in the 2 and 7 o'clock positions and the piston is worn unevenly on the thrust faces, chances are that rod is bent. A rod exhibiting these conditions should be taken to the machine shop for close checking. Tell the machinist what you suspect. He may be able to correct it by straightening the rod. If he can't, you'll have to replace it.

Out-of-Round & Size—The first thing an engine machinist checks is for an out-of-round and over or undersize bearing bore. To obtain the proper bearing crush, a bearing bore must be slightly smaller than the OD of the bearing halves. The slight undersize of the bore causes the bearing halves to *crush,* which locks them in place and makes the bearing conform

to the shape of the bore. Bearings also depend on this tight fit to transfer heat from the bearing. Read about bearing crush in the sidebar on opposite page.

This check is done with a special dial indicator. Calibrated with a micrometer, such as a dial bore gage, the dial indicator indicates *variance from the standard.* When a rod—less bearings, but with torqued nuts—is placed over the indicator, it reads 0 if the bore is to spec. The needle stays on 0 or within 0.0002 in. as the rod is rotated. If the bore is out-of-round, the dial reads +X.XXXX for too large a bore or −X.XXXX in. when it's too small. Connecting-rod-bearing bores should be 2.3252−2.3247 in., or 2.3250 ±0.0002 in.

Before delivering the rods to the machinist, remove the bearings and torque the nuts. Clean the cap-to-rod mating surfaces before reinstalling the cap. Double-check that the cap and rod numbers are the same and are on the same side. If you don't have a *rod vise*—a precision vise for clamping connecting rods—clamp the rod in a vise between two blocks of wood to torque the nuts. Support the rods using one of these methods to prevent rod damage.

Oil the bolt threads, install the nuts and torque those with 3/8-in. bolts 50 ft-lb. Torque those with 7/16-in. bolts 67 ft-lb. Do this and you won't have to pay the machinist to do it.

Recondition Rods—If the big end is out-of-round or over or undersize, it

will have to be reconditioned, or *resized,* by honing.

Reconditioning, or sizing, a rod begins with removing the cap. A small amount of metal is then precision ground from the cap parting surfaces. Afterwards, the cap is reinstalled and nuts torqued to spec. Obviously the bearing bore is very out-of-round now. It is smaller top to bottom, but virtually the same across the parting line.

The big end is then honed, usually two rods at a time as shown. Honing rods in pairs helps keep them square to the hone, thus giving a truer bearing bore. Reconditioning also removes any taper from the bore.

Rod Bolts—Rod bolts have not been a problem in street-use big-blocks. If your engine has seen nothing more than regular trips to work, the supermarket and a two-week vacation on the interstates once a year, I don't see any reason to get concerned and check the rod bolts. High rpm kills rod bolts, so if you've been "twisting" your big-block on cruise night, make a point of inspecting the rod bolts. If you have not owned the engine since it was new, check the rod bolts.

You'll have to remove the bolts from the rods to inspect them. Be aware that when one or both rod bolts are removed from a rod and replaced, the big end must be checked and probably reconditioned. The reason for this is the bolts act as cap-locating dowels. Once a bolt is removed and

An arbor press or hydraulic press must be used to disassemble a connecting rod and piston with pressed-in pins. **NEVER** hammer on a piston pin to remove it. You'll only destroy the pin, piston and possibly the rod. To remove a floating pin, remove one retaining ring and slide the pin out.

Expanding the small end of one connecting rod by heating in electric heater while a rod and piston are assembled. Although pressed-in pins can be installed as they were removed—with a press—heating the rod is the best method. After heating, the pin will slide into the rod without the danger of galling it or the rod. As the rod cools, the small end contracts and locks the pin in place.

replaced, the cap will be slightly out of position.

And there's more to removing a rod bolt than overhanding the old ball-peen. First, clamp the rod in a rod vise or between wooden blocks in a conventional vise. Remove one of the nuts and hammer that bolt out of its bore with a punch—a brass punch is preferable.

Inspect the bolt under a bright light; look for cracks. Use a magnifying glass if you have one. Before attempting to remove the other rod bolt, reinstall the one you just took out, or put in a new one.

To install the bolt, clamp the rod using the same method you used before. Insert the bolt in the rod and rotate its head so it will fit the spotface on the rod. Tap it into place.

As an aside, it is a good idea to replace the rod-bolt nuts, especially if they were removed with an impact wrench. Buy new nuts now so you'll have them when you are ready to assemble the engine.

PISTON-AND-ROD DISASSEMBLY & REASSEMBLY

If your engine has pressed-in pins, this section is directed at you. If floating pins are used, skip this section.

A press with the appropriate mandrels is required to remove pressed-in pins. If you have a press and the proper tooling, by all means do the job and save yourself some money. But if you don't, my advice is to have your engine machinist do the removal-and-replacement (R&R) work.

Above all, *do not hammer on a piston pin!* Even though you may get them apart, you'll ruin the pin, piston and rod in the process. This applies to power tools as well. Trying to blast a press-fit pin out with an impact, or *zip,* gun is an automotive crime of the first degree. *Always* use a press.

Remove Pin—To remove a piston pin correctly, rest the piston on a hollow mandrel, squarely on its pin boss. The hole must be large enough to allow the pin to pass through. There must also be a hole in the press bed for the same reason. A punch must be used between the press ram and the pin. It is best to use a *shouldered* punch—one that butts against the pin end and has a reduced diameter to center the punch in the pin. Center the pin, mandrel and punch under the press ram. Before you start to press out the pin, make doubly sure the pin boss is resting squarely on the mandrel. Now push the pin out. There'll be a loud *bang* the instant the pin breaks loose.

Install Pin—Assembling the pin and piston can be done two ways: with a press or with heat. Pressing a pin is easy enough to understand; the pin is forced into a bore that is smaller than it is. However, when the small end of the rod is heated, the bore expands, allowing the pin to slide in by hand. When the rod cools, it clamps tightly on the pin.

Pressing in a pin is basically the reverse of pressing it out, but you must be extra careful to line up pin, piston and rod. First, the rod and piston must be assembled in their correct relationship to one another. And a rod can be easily ruined if the pin is not aligned with the rod pin bore. It won't do the piston any good either. A little lubricant on the pin—moly will do—reduces the chance of galling. It also reduces the force needed to press in the pin. Finally, the pin should be installed so it projects equally from both sides of the rod.

Using heat to install piston pins ensures against galling. To use the heat method, a heat source such as an acetylene or propane torch is needed. It is very difficult to monitor the amount of heat applied to the rod, or how far in to install the pin—you have to be *quick.* If you take too much time, the rod will cool and clamp onto the pin. A press is required to move it any farther. If this is your first time using heat to assemble rods and pistons, count on finishing about half of them with a press. Leave this job to the experienced hand.

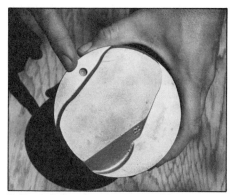

Dot on piston goes to the outside, opposite intake-valve notch. Connecting-rod squirt hole goes toward center of engine. When assembling a rod and piston, make sure the dot and squirt hole are on opposite sides.

Groove (arrow) in rod cap is for camshaft and cylinder-wall lubrication. Oil passage should be on the camshaft side of the block. Rod number is on opposite side.

Rod-and-piston assemblies stored relative to their installed positions: rod numbers and piston locating dots to the outside, oil passages to the inside. Intake-valve clearance notch to the inside.

If you're building a big-block for another 100,000 miles, a Cloyes True Roller timing set is a good choice. Chain is wider than stock and features low-friction rollers. Additionally, crank sprocket has three keyway slots to vary cam timing: stock, advanced for power at lower rpm and retarded for power at higher rpm.

Well-equipped machine shops have a special rod-heating device that does a beautiful job of heating connecting rods. If you want the best possible piston/rod assembly, look for a machine shop with a rod heater—or a machinist who is proficient with a torch.

Heat the small end of the rod cherry red, no more. Considerable practice is the only insurance of a 100% success rate for installing each pin before the rod grabs it. Slide the pin in so each end is the same distance from its side of the piston when the rod is centered in the piston.

No matter which method is used, make doubly sure the pistons are installed correctly on their rods. Big-block-Chevy piston bores are offset 0.030 in. to the *right*—the pin is 0.060 in. closer to the right skirt than the left. The purpose of this offset is to *preload* the piston against the right cylinder wall to reduce piston-slap.

If you install the piston 180° out of position, you'll have to push the pin out and turn the rod or piston around. But how do you distinguish the right side of the piston from the left?

Pistons with pop-ups give you a clue because of the dome shape. A pop-up must conform to the shape of the combustion chamber. But Chevy had a better idea. Look at the piston dome. You'll find a dot—it's small. This dot must point in the same direction as the connecting-rod number. It points to the right on pistons that install in the right cylinder bank and to the left on those that go in the left. The same follows for the rod numbers. Numbers are opposite the oil squirt hole—if your rods have them. Dot and number must point in the direction of their cylinder bank when installed in the engine—right-bank piston dots point to the right, and vice versa.

One more point to keep in mind with unnumbered rods: Yes, the squirt holes do point to the opposite cylinder bank. But that's only half of the story. The rod face with the big bearing-bore chamfer installs toward the bearing-journal radius and away from the rod it is paired with. This is a must because if the rod was installed in the reverse direction, the rod bearing would be forced into the journal radius, making the bearing edge-ride.

So, when the piston-rod assemblies are correctly installed, both the dot and the numbers point toward the outside of the engine, the oil squirt hole points to the opposite bank, and the rod chamfer goes toward the journal radius.

When you've finished assembling the pistons and rods, lay the assemblies out as if they were in the engine, as shown above. Rods 1, 3, 5 and 7 will be to the right—front to rear; 2, 4, 6 and 8 will be to your left. Both the piston dots and rod numerals should be facing the outside of the block.

TIMING CHAIN & SPROCKETS

While we are working with bottom-end components, let's consider timing-chain wear. You already know about the wear characteristics of timing chains and sprockets, and how to check them for wear, pages 56—57. What you need to know is how to get the most for your timing-chain dollar.

Of the three sets of timing chains and gears offered by Chevy, you'll want the 3/4-in.-wide Link-Belt design. The most durable and practical cam-drive combination, it was the stock offering for all big-blocks in 1965 and '66. In 1967 the same type of chain and nylon gear was retained, but the width was reduced to 5/8 in. The earlier, wider chain-and-sprocket setup will interchange with any big-block chain-and-sprocket set.

Some truck engines are fitted with a double-row "roller" chain and all-steel sprockets. This offering is stronger than the nylon-toothed pair, but somewhat noisier. Although this original-equipment truck chain is not a real roller, there are no nylon sprocket teeth to deteriorate or break.

If you want the most durable cam-drive setup, the Cloyes True Roller chain-and-sprocket set is for you.

Unthread the four cover bolts and remove pump cover. Hold pump so pickup intake is up so gears won't fall out.

Before removing them, mark the gears so they can be installed in the same relative position. Use a felt-tip marker or electric etching tool to mark the gears, not a punch. Turn the pump over and let the gears drop into the palm of your hand.

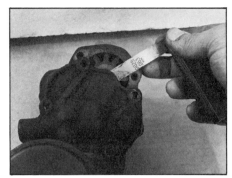

Measure gear-to-housing clearance with feeler gages. If clearance exceeds 0.0025 in., lap pump body. If clearance is less, lap gears.

Sprockets are all-steel and the individual chain links feature honest-to-Pete rollers. Additionally, the crank sprocket has two extra keyways for tailoring cam timing to engine application—advance cam 4° for torque at the low-end, or retard it 4° for high-rpm power. This setup virtually eliminates chain-elongation worries and keeps timing variation to an absolute minumum. The cam sprocket is also balanced—important if your big-block will be run at high rpm.

OIL PUMP & DRIVE SHAFT

The big-block oil pump has earned itself a bulletproof reputation. Unless your engine has digested large dirt or metal particles, or you ran the same oil for the last six years, the oil pump should be OK. Certainly this doesn't mean an inspection isn't necessary, only that the pump design is sound and problems are few.

A few words of advice: Use a stock oil pump. If the original pump is OK, reuse it. If not, replace it with an equivalent pump. A machine shop may try to sell you a high-volume oil pump, but don't bother. There are a couple of things you can do to the standard pump to improve it for high-performance use. I don't go into these here. But if modifying the oil pump is something you want to investigate, read HPBooks' *How to Hotrod Big-block Chevys.* In addition to oil-pump modifications, there are hundreds of tips for the high-performance enthusiast between those covers.

Back to the stock pump. Scrub the outside of the pump with a stiff-bristle brush and solvent. This will keep grit out of the pump when you open it up. Submerge the pickup tube in a fresh pan of solvent and rotate the pump driveshaft *clockwise* until solvent gushes from the output hole. Pretty soon the solvent will run clean from the pump innards and you can stop turning.

Disassemble & Inspect Pump—Turn the pump upside down—watch for draining solvent—and undo the five cover screws. Don't let the gears fall out onto the floor. This will damage them. Mark the gears, as shown, before removing them—they should be reinstalled in the same relative position.

Visually inspect the pump body and gears. Deep scoring of the pump body and chipped gear teeth indicate that large dirt or metal particles passed through the pump—get a new pump. Minor scratching and scoring from dirty oil is normal—reuse the pump.

End Clearance—Most important of the pump checks is *end clearance*—distance between the gears and the pump cover. Measure end clearance with feeler gages. Hold the cover firmly over the pump body so it partially covers both gears. Find the feeler gage that fits between gears and cover. Stock-pump end clearance seems to vary from 0.003 to 0.004 in.

You can also check end clearance with a depth gage or Plastigage. The depth gage will give the most accurate readings, but a depth gage isn't found in every tool box. Plastigage is used in a manner similar to checking bearing clearance, page 116.

Chances are end clearance is close to normal if you kept the oil clean. Even so, consider reducing rotor-to-cover end clearance. The big-block-Chevy oil pump works best with a 0.0025-in. end clearance. It will prime quicker and pump more efficiently.

End clearance is reduced by *lapping* the pump body on abrasive paper backed by a flat, rigid surface—a thick piece of glass works well. To lap the body, buy one sheet of No. 220 Wet-or-Dry emery paper. Find a hefty piece of glass—a glass shop might have a broken piece you could use. The surface supporting the paper must be *very* flat and thick enough not to break or bend when you press on it with the pump body. Lay the paper down grit-side up. Lightly flush the paper with water. Place the pump body—cover-side down and without gears—on the paper.

Move the pump back and forth with *light pressure.* Don't bear down on the pump or you'll remove too much material. After several passes, turn the pump 90° and continue. Keep turning the pump so the lapping marks go in different directions. Don't lap too long, it doesn't take much to remove a thousandth or two.

Make frequent feeler-gage checks. You should be able to get end clearance right at 0.0025 in. Of course, it is better to have end clearance slightly over 0.0025 in. than it is to end up with less. If you overdo it, lap the gears just like you lapped the body until end clearance is 0.0025 in.

When you reinstall the gears to check end clearance, be sure to wipe any grit from the pump-body mating surfaces. If you don't, the tiny granules will affect your measurements and you'll remove too much material.

After lapping, clean the pump in solvent. Remove every last particle of lapping grit. You don't want that junk

Dirty oil takes its toll on all parts of an engine, particularly the oil pump—the pump is *ahead* of the oil filter. Pump cover is from the dirty engine pictured in the teardown chapter. To remove light scratches from a pump cover, lap it face down on 220-grit emery paper on a flat surface.

Lapping grit or dirt must be washed from the pump before reassembly. Flood the pump body with solvent.

pumped through the lubrication system of your newly rebuilt engine.

Pressure-Relief Valve—There isn't much to service here. About the only thing you need to check is that the valve moves freely. Use a small pick or screwdriver to do this. The relief valve is accessible outside of the oil-pump cover, right below the pickup. The relief spring butts against a washer, which is secured by a pin. If you can push the washer in against spring pressure, the relief valve is OK.

KNOCK, KNOCK, PING, PING

We've all heard the death rattle of yesterday's engines running on today's gasoline. A constant, cacophonous dirge of pounding pistons and ringing blocks accompanies every downward stroke of the accelerator. Even newer low-compression engines ping and run on, not to mention older big-blocks and their once "low compression" of 10.25:1! Internal elements cry audibly for mercy with each power stroke. So what can you do now that high-octane gasoline is but a memory?

There's a lot to do. Which anti-knock cure is best depends on your engine. Later engines are supposed to run well on modern gasoline, so getting all the internal parts in shape should do the trick. The key words are "supposed to," so keep reading. Earlier engines may require extensive modification before they will run well on low-octane gasoline.

The first decision; will the engine be run on leaded or unleaded fuel? If your engine is a 1975 or later model it doesn't matter, it will run on unleaded fuel. Earlier engines were designed for leaded premium fuels, which are fast disappearing.

To make the switch to unleaded fuel, you'll need to install *stellite* valve seats and bronze valve guides or inserts. Without these two modifications the valves will suffer from lack of lubrication that was provided by leaded fuel.

Your big-block may be able to run with 93-octane unleaded without pinging, but the valves will be "running dry." They will soon wear out. Hard seats and bronze guides should also be used if your big-block is going to be run on propane fuel. TRW markets high-nickel valves that withstand heat better than stock replacements, so you may want to try them.

Buying new seats and guides is expensive, and so is rigging a propane fuel system. Leaded regular fuel is here to stay, apparently, so consider modifying your big-block to run on it. Simply installing two head gaskets will cheaply and effectively reduce any engine's compression ratio so it will run on regular—especially if it's a later version with stock low compression.

Another approach is to install low-compression pistons and heads. Open-chamber heads on a "closed-chamber" block can lower compression to the 7.5:1

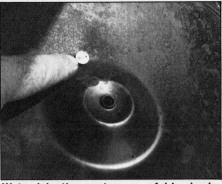

Water-injection systems are fairly simple. This single emitter, fitted to the underside of the air filter lid, sprays the water into a fine mist.

range—low enough to run well on regular fuel. However, such a modification will reduce engine power output. Juggling pistons and heads can get your engine into the 9.0:1 bracket, which seems the best compromise between anti-knock and power loss.

Consider the alternatives outlined in Chapter 3. You may be able to assemble an engine with the compression you want. Also consider the TRW catalog. TRW's low-compression Turbo pistons can be used—and you don't have to change heads.

When changing pistons, keep in mind that flat-top designs impede the flame front less than domed pistons. A faster, more even flame front is less apt to permit detonation or preignition, so flat-top pistons are the good guys in the anti-knock battle. Also, moly rings have been found to be equal to chrome rings under detonation, so keep on using moly rings. Fairly tight ring end gaps help; about 0.014 in. is good.

However, the trump card when it comes to knock-busting is water injection. There are many units on the market, but the principle is the same for all. By injecting water into the fuel charge, combustion temperatures are reduced enough to prevent detonation. Carbon buildup in the combustion chamber is eliminated, further reducing detonation and preignition.

Parts swapping is not necessary and power is maintained because you're not lowering compression to eliminate detonation. Compression below 11:1 seems to work fine with water injection. So unless you're restoring an LS-6 or L-88, you should have no problems. Even with these high-compression engines, detonation and preignition should be minimized with water injection.

All you need to do is to rebuild your engine to its original specs, then install a water-injection system. Maintenance is confined to keeping the reservoir full of water *during warm weather.*

When the mercury approaches freezing, a water/alcohol mixture must be used. This is no big deal unless freezing occurs at the carburetor, causing the throttle to jam. Read the manufacturer's instructions that accompanied the water-injection kit for the proper use of alcohol—methanol in this case.

You'll notice the increase in usable horsepower, especially when towing. Long, heavy pulls go a lot faster when you don't have to lift off the throttle to prevent knocking. You can also run your 10.25:1 closed-chamber powerhouse on almost any fuel you find, if necessary. Keep in mind, though, that water injection isn't a total cure-all. If you burn unleaded you'll still have the valve-lubrication problem, but your engine will run better than it would otherwise.

Combine water injection with a spark control, available in the aftermarket, and you have a setup that can effectively pump 10 extra octane points into your gasoline. Spark-control devices are available from Carter and MSD (Autotronics). Get information on these products from your local speed-equipment store. One of these units would be just what the doctor ordered for your L-88 Corvette.

The choice of detonation suppressor also depends on your driving habits. If you want the most from your big-block you'll also want the most detonation suppression. Towing, mountain driving and traveling with full loads all demand the most from any engine. On the other hand you may use your car for little more than short trips around town and therefore don't need much detonation suppression, if any.

For light-duty, fitting two head gaskets under each head should be just right. However, tailoring your own combustion chamber by changing pistons or heads is more effective than the extra head gaskets. It also gives you the opportunity to install a better breathing head to recoup some of your power losses. The most effective anti-detonation device is water injection. And if you add up the cost of tracking down parts and getting them into shape, it may be considerably cheaper.

6 Cylinder-Head Reconditioning

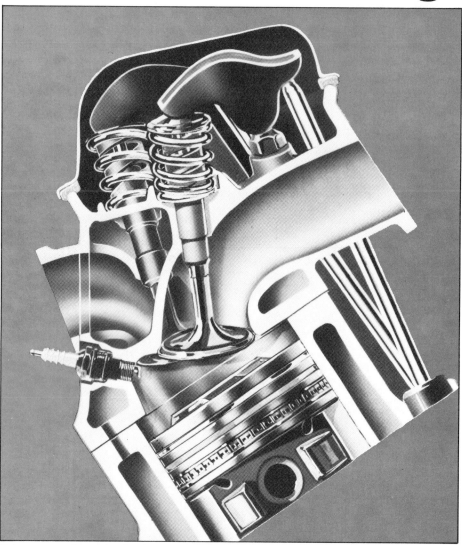

Unique procupine valve layout of Mark IV Chevy gave excellent breathing by reducing curvature of ports and valve shrouding. Note separate valve guides. Photo courtesy of Chevrolet.

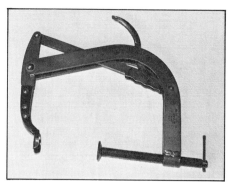

Valve assembly and disassembly requires a valve-spring compressor. If you don't have one, you can rent one. Screw adjustment governs how far spring compresses. Don't compress a valve spring more than what is required to remove keepers.

All cylinder-head teardown, inspection, reconditioning and assembly work is covered in one chapter because this is how it's done in an engine shop. And once head work begins, it continues to the point at which the head or heads are ready to install on the engine.

There are several good reasons why you should farm out your cylinder-head work. First, there's the equipment you'll need: valve grinder, valve-seat grinder, and valve-spring compressor and tester. These tools are so specialized and expensive that no one rents them. And you sure can't justify buying them for doing one engine job.

Second, even if you had access to a machine shop full of head-reconditioning tools, you would still have to learn how to use them. It's one thing to say "grind the valves" and something else to do it.

So, if you don't have the tools or skills, deliver both heads to a reputable machine shop and have them do the job. It won't take long and, if you consider the time saved, the job doesn't cost that much.

A word to the wise; don't shop for a machine shop like you would for gas. Price is not the main consideration. It usually pays to go to the somewhat more expensive but faster, cleaner and more professional shop. The extra care and time spent on *your* engine costs money—so don't worry about spending a little extra now. Better a little now than a whole lot later down the road.

To make the first stages of cylinder-head work more pleasant, clean them. Loosen the crud with engine degreaser, then wash off the heads with a garden hose. Or, if you don't like making a mess at home, use the local car wash. Spray the heads with degreaser and blast off the dirt with the pressure spray.

DISASSEMBLY
Rocker-and-Ball Relationship—Because of its importance, I'll repeat this warning: **Keep the rocker arms and**

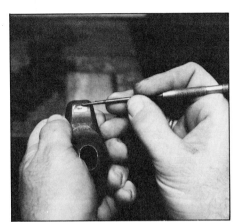

Worn rocker-arm tip means rocker-arm replacement. Although a rocker tip can be ground smooth, the hardened surface will be removed. Rapid tip wear will result.

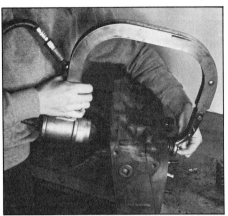

Air-powered spring compressor being used to disassemble heads. If spring will not compress, tap retainer lightly with a mallet to break it loose from the keepers. Better yet, use the socket-and-hammer method discussed in text to break retainers loose before you use the spring compressor.

their balls together. Wire them together so there's no chance of mixing them up. They are just like camshafts and lifters; once run, they wear-in and must be kept as matched sets. Otherwise the rockers and balls will gall and valve action will suffer. Not only will you have to replace the rocker arms and balls, metal particles will be circulated through the oil pump.

Valve Removal—The first step in head reconditioning calls for the first specialized tool—a *valve-spring compressor.* The most common type is the *C-type,* so called because it looks like a huge C-clamp. It is designed for use with the head off of the block. A *fork-type* compressor is best used with installed heads.

A C-type compressor fits around the head. One end butts against the valve head, the other end straddles the valve-spring retainer. A lever on the compressor moves its two jaws together with an overcenter effect. This compresses the spring, forcing the spring retainer down the valve stem to expose the two *keepers, collets, locks* or *keys*—whatever you prefer calling them. Their purpose is

to lock the spring retainer to the valve stem with a wedging action, so the retainer holds the valve spring in compression.

A common problem when using a spring compressor is breaking the retainer loose from its keepers. If you try to force the compressor, you just bend its frame. To prevent this, there are two methods you can use. First, lever down with the compressor and tap the retainer/compressor with a soft mallet. This will break the retainer loose from its keepers, allowing the spring to be compressed.

Be careful when doing this; parts tend to go flying. Keep your head and body out of the way, but watch the keepers' trajectories. Find them before continuing.

An alternate method of breaking keepers loose is to lay the head flat on the bench with a small piece of wood in the combustion chamber, against the valve heads. Select a socket that fits squarely against the retainer, but allows sufficient clearance around the valve stem and keepers. A deep 3/4 or 1-in. socket will do.

Set the hex-end of the socket against the spring retainer and give it

a sharp blow with a mallet. Keep a firm grip on the socket with clamping pliers, or Vise Grips, as the keepers may fly off, freeing the retainer and spring. Stand at arm's length, out of line of the retainer and spring just in case this happens.

If they fly off, the socket will contain the keepers so they won't be launched into the fourth dimension. As a bonus, the spring and retainer will be free so you won't need a spring compressor. This is the quick-and-dirty method of spring disassembly. Using a spring compressor is better, although slower. You'll need one for reassembly, anyway. Remember, you're a mechanic, not a blacksmith. Go easy.

There's no need to keep the keepers and retainers in order, just dump them into a container so they won't get lost. You must keep the valves, springs and any shims in order, however. The best way is to use a 2x4 or piece of cardboard, just as you did with the pushrods. Insert each valve stem through a hole and slip its spring over the stem.

Put 16 holes in the wood or cardboard and number them 1 through 8 from left to right for each head, viewing the head from the exhaust ports. Indicate which head the valves were installed in and mark each head likewise. Parts are a lot harder to mix up when each one has its own special spot—and is identified.

Some engines use valves springs with *dampers*—flat wire wound to fit tightly inside each spring. Similar to a suspension shock absorber, the valve-spring damper controls, or *damps,* unwanted spring movement. Leave each damper inside its spring so they will not get separated.

Dual springs—one smaller spring inside the main spring—are found on some pre-'70 427s and almost all big-blocks after that year. Handle dual springs like dampers—keep them paired with their outer springs.

When you remove each spring from its pocket, you'll find a hardened-steel shim underneath.

These shims protect the head against chafing from the high-load springs. If they are left out, the bottom of the spring will eat into the head—a very real problem with aluminum heads.

In addition to the factory-installed shims, you may find other shims under the springs if a valve job was performed on the heads. These shims correct certain valve-spring conditions, which I'll discuss later.

After all the springs are removed, place a flat file against the edge of each valve tip and rotate the valve a few times. This will remove the slight burr that can cause a valve to hang up in its guide as you remove it.

Pull the valve and it should slide through its seal and out the guide. If the valve hangs up in the seal, pull the seal off the valve stem with a pair of pliers. Throw the seal away; new ones are included in the engine gasket set. Many high-performance big-blocks have the oil seal attached to the bottom of the retainer—don't remove it.

Put the valves in order and store the springs and shims before someone bumps into them and knocks them out of order.

Valve Rotators—If you are working with a truck engine, either a 366 or 427, the exhaust valves use *rotators* under the springs that turn the exhaust valves as they open and close. This ensures that the valve closes in a different spot on its seat to prevent burning by promoting even wear and cooler valve operation. Even though valves rotate slightly with conventional retainers, rotators turn them positively and faster. All you have to do is clean the rotators and store them so they don't get misplaced.

Rocker Studs & Pushrod Guides—The cylinder heads should now be bare except for valve guides, pushrod guides and rocker-arm studs. Barring unusual circumstances, there is no service work to do with either the studs or pushrod guides.

Wear on rocker-arm studs and pushrod guides is minimal and both last indefinitely in normal service. About all you need to do is check the rocker studs for tightness; they should be torqued 50 ft-lb. Use a 9/16-in. deep socket with your torque wrench.

If you find a loose stud, remove it. Inspect the threads *before* attempting to torque it to spec. Make sure the

After disassembling heads, scrape off all old gasket material. If they are cast iron, hot-tanking will make this job much easier.

Head was hot-tanked, then the combustion chambers bead-blasted. Stubborn carbon in ports is being removed with a scraper. A lot of time goes into reconditioning a pair of cylinder heads, so don't grumble about the bill!

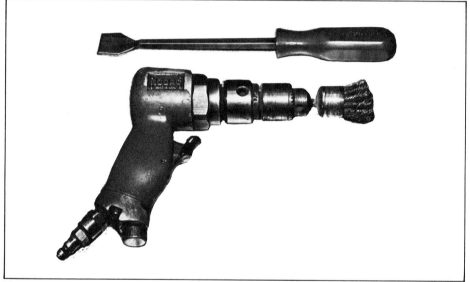

Two tools that make cylinder-head cleaning much easier: gasket scraper and air- or electric-powered wire brush. You've got to see what you're doing during a valve job, so remove the carbon from around the valve seats.

threads on the stud or in the hole are not stripped. If they are OK, reinstall the stud with a dab or two of red Loctite on the threads. This will keep the stud from loosening again.

Cleaning—The easiest way to clean cylinder heads is in a hot tank—unless they are aluminum. If you haven't shipped the block and its parts to the machine shop, send the heads along to have them tanked at the same time. Otherwise, follow up your degreasing job with a solvent bath.

Don't worry about carbon in the combustion chambers. It is too baked-on for the solvent to do any good. Knock the largest carbon pieces off with a blunt screwdriver or chisel, but be careful not to nick the valve seats, sparkplug threads and head-gasket surface.

Finish the job with a wire brush. An electric drill with a cup-type wire brush works well. Get in all the corners and don't be shy about running the brush into the valve seats. Employ the gasket scraper on the gasket surfaces: head-to-block and intake- and exhaust-manifold surfaces.

Be extremely careful while scraping the head-gasket surface. Gouge it and you'll provide a path for combustion pressure to blow past the gasket.

The edges of all the water passages need cleaning too. Use a pocketknife or dull screwdriver to remove the large deposits, then finish up with a few passes of a rat-tail file. Don't remove any metal—just the scale. To remove the junk that fell into the water passages, hold the head right-side up over a trash can and move it

If head-gasket surface is damaged or warped more that 0.004 in. over its length, the cure is milling. Most engine shops don't bother checking for warpage, they simply mill the heads until the gasket surfaces clean up. However, warpage is not a problem with big-block Chevy heads, so you shouldn't automatically spring for having them milled. Check them to make sure.

METAL-REMOVAL CHART (IN.)		
	INTAKE MANIFOLD	
Cylinder Head	Side	Bottom
0.005	0.004	0.010
0.010	0.007	0.019
0.015	0.011	0.029
0.020	0.014	0.038
0.025	0.018	0.048
0.030	0.021	0.058

Chart tells how much to remove from intake-manifold bottom and sides to correspond to what was milled from heads.

around until it all falls out. Just don't lose your grip and drop the head on your foot!

Wire brush the valves, removing all traces of carbon. This could take a while, so be patient. A wire wheel on a bench grinder works best, but the electric-drill/wire-brush attachment will do if you can hold the drill or each valve stationary.

Head Warpage—Some engines warp heads with the regularity of a Swiss watch, but not the big-block Chevy. Those massive heads can take a lot of punishment before they take on a curved shape. If you haven't overheated your big-block and it doesn't have 200,000 miles on the clock, the heads should be flat. Repeated hot/cold cycling after many, many miles or frequent overheating will warp a head—even a big-block Chevy. Regardless, check the heads for warpage. To make this check you'll need your feeler gages and a precision straightedge such as the one for checking the block, page 71.

Before checking warpage, make sure the straightedge and the cylinder-head gasket surface are perfectly clean: no gasket material, dirt or whatever. If they aren't, you'll get bogus readings. Wipe both surfaces down thoroughly before taking any measurements.

Make the first check with the straightedge lengthwise on the head. Try to pass a 0.004-in. feeler gage under it. If it fits, pull the feeler gage out and try. thicker ones. When you find one that fits the gap snugly, you've found the amount of warp in

Example: If a 0.007-in. gage fits, the head is warped 0.007 in.

After checking the head lengthwise in different positions across the head, make numerous checks with the straightedge placed diagonally.

What's good and what's bad depends on the type of head gasket you plan on using. Most head gaskets are of the *composition-type* as opposed to *shim-type* gaskets. A composition gasket has several layers of material; usually asbestos and aluminum or copper. These layers allow the gasket to conform to slight irregularities. For instance, warpage up to 0.006 in. poses no problem with a composition gasket.

A shim gasket, on the other hand, is made from a single layer of steel, copper or aluminum. That single thickness of light-gage metal is not capable of conforming to even slight irregularities; cylinder-head gasket surfaces must be flawless. So if you plan on using shim-type gaskets, head warpage must be kept to a minimum. Anything more than 0.003 in. is too much.

In case you were wondering about which gasket to use, the composition gasket requires a follow-up torquing after the engine has been warmed. A shim gasket doesn't need the second torque job. If your engine is covered with power accessories that block access to all the head bolts you may want to consider using the shim gasket, otherwise use the composition gasket.

Cylinder-Head Milling—If one or both heads are warped, *both* must be milled flat. The same amount must be removed from both heads so both engine banks will have the same compression ratio and so the intake manifold will fit properly. Mill only one head and you'll end up with a rough idling engine. Something else to consider: If more than 0.030 in. is removed from the heads, the sides and

bottom of the intake manifold must be milled as well. This is because the heads will sit lower on the block, causing misalignment between the ports and bolt holes in the heads and manifold. Refer to the chart for intake-manifold milling.

Milling Tips—Make an effort to limit cylinder-head milling to 0.015 in.; no more. You will do yourself a couple of favors. First, you will not be raising the compression ratio to the point where it will be impossible for your newly rebuilt engine to run on modern gasoline. It doesn't take much milling before you get into detonation territory, especially with older, closed-chamber heads. Remember, these engines started with a minimum of 10.25:1 compression!

Typical values are 1/10 of a ratio for 0.030 in. off the head, or 1/4 of a ratio for 0.060 in. If your early 396 started with 10.25:1 and you *cut* the closed-chamber heads 0.060 in., your engine could have a 10.50:1 compression ratio!

Second, if you stay within a 0.015-in. milling limit, you won't have to concern yourself with milling the manifold. So if it looks like you will need to mill more than 0.015 in., buy another head with the money you would have spent on machining the manifold. You'll also avoid the problems associated with a high-compression engine in a low-compression era.

VALVE GUIDES & STEMS

Back in Chapter 3 I mentioned the problems with the big-block valve guides. To refresh your memory, big-block heads have separate valve guides. What's significant here is the holes the valve guides are pressed into are not concentric with the valve seats. Consequently, replacing the guides is not a simple matter of driving out the old and driving in the new. New guides can be installed and many

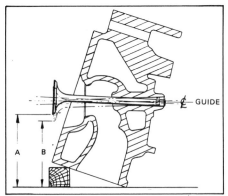

Wiggling valve in its guide is the time-honored method of judging guide wear. Insert valve in its guide so its tip is flush with the rocker-arm end of the guide. Grasp valve by the head and wiggle it up and down. Measure and figure valve-head movement, or A - B. Divide result by 3.5 to get *approximate* stem-to-guide clearance. Although maximum movement is usually found in a vertical plane as shown, check valve movement in other directions to make sure. Drawing by Tom Monroe.

VALVE-STEM-TO-GUIDE CLEARANCE		
	Intake (in.)	Exhaust (in.)
Standard	0.0010—0.0027	0.0010—0.0027
Pre-'68 427	0.0010—0.0027	0.0015—0.0032
Pre-'73 454*	0.0010—0.0037	0.0010—0.0047
*Solid-lifter cam		

manufacturers offer the parts, but not all machine shops like to handle the job because of the concentricity problem.

We've come to an interesting point. There are several methods of handling big-block guides and different machine shops have their preferences. Machine-shop A will replace the guides with new ones, machine-shop B will only install guide inserts while machine-shop C will only use oversize valve stems. It depends on where you live, what tools are available in the shops and the machinist's ability and preference. I talk in detail about each of these methods later.

For street-driven engines the advantages and disadvantages of most fixes are academic. So searching for the hot-tip in valve-guide technology is next to a waste of time. Read about the different approaches available, pick out the one you like and find the machine shop that does it the way you feel is best for your application. Trying to persuade your machinist to use a method he is not comfortable with won't make you a friend and will likely inflate the bill.

Valve-Guide Wear—Let's consider guide wear and how to measure it.

As a rocker-arm pushes its valve open, it also forces the valve stem against the guide. This action wears both the guide and stem. Keep in mind when someone says, "The guides are worn out," they really mean the clearance between guides and stems is excessive—both the stem and guide could be at fault.

Let's start with the valve guides because they wear most. A quick and pretty accurate wear check uses a valve in the guide. Insert the valve in the guide so the tip of the stem is flush with the top of the guide.

While holding the valve by the head, rock the valve back and forth. Move it in different directions until you find the one that gives the most movement. Then measure the movement, or *play,* with a 6-in. scale or the depth-gage end of vernier calipers, or a dial indicator if you want a really precise reading. Divide this measurement by 3.5 and you'll get *approximate stem-to-guide clearance.* This method still does not tell you if the valve stem or guide is the culprit, but you can usually count on the guide being worn.

Measure Valve-Stem Diameter—To be positive about it you'll have to mike the valve stem with a 1-in. micrometer as explained on pages 100 and 101. Intake-valve stems should measure 0.3715—0.3722 in. up to '69. After that Chevy just lists them as having a 0.3719-in. diameter. Exhaust-valve stems are the same. Until '69 the range was 0.3713—0.3720 in. Post-'69 exhaust-valve stems are 0.3717 in.

Direct Measurement—To get the most accurate guide-wear measurement you'll need a very small telescoping gage or *small-hole gage*—sometimes called a *ball gage*—and a 1-in. micrometer. A small-hole *C-gage* is the size needed. Insert the gage into the guide to find the point of maximum wear. As you expand the gage, rotate it and run it up and down until you find the widest point in the guide. Maximum wear occurs at one end of the guide or the other. Remove the gage and mike it. Standard guide diameter less this figure is *exact* guide wear.

How Much Guide Clearance—Find stem-to-guide clearance by subtract-

It's appropriate that knurling valve guides doesn't take long; they don't last long. Knurling tool is driven into guide, then run through with a drill motor.

ing minimum valve-stem diameter from maximum guide diameter. Nearly all big-blocks have the same range of acceptable guide clearance, 0.0010—0.0027 in. Exceptions are the pre-'68 high-performance 427s. They have a looser 0.0015—0.0032-in. exhaust-valve-guide clearance; intake stem-to-guide clearance is the same as standard big-blocks. Pre-'73 mechanical-cam 454s use a wider range on intakes, 0.0010—0.0037 in. and 0.0010—0.0047 in. for exhausts. If stem-to-guide clearance exceeds these dimensions you need to restore the guide—and maybe the valve stem—to factory specs. Stem-to-guide clearances are in the table above.

Valve-Guide Reconditioning—If you find too much wear, there are several methods of restoring valve guides. The methods I prefer retain the original guides and, consequently, do not require guide replacement. Thus, the guide-to-valve-seat concentricity problem is avoided. As previously discussed, methods requiring new guides may cause concentricity headaches.

To make life easier, let's start with the methods that don't involve guide replacement.

Guide Knurling—The cheapest way to restore stem-to-guide clearance is by *knurling* the guides. Just like other cheap methods, knurling is a stop-gap

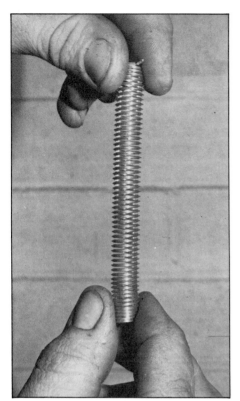

Screw-in guide insert threads into tapped guide. After installation, insert is trimmed, locked and reamed to size. Once in great favor with rebuilders, this type of guide insert has fallen into disuse. Improperly installed it can actually pump oil into the port.

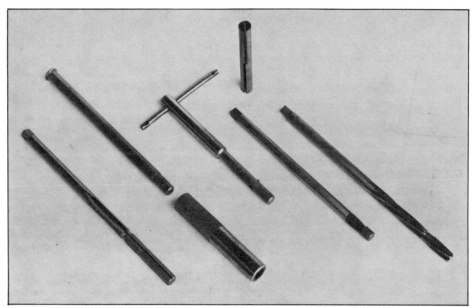

Thin-wall bronze valve-guide insert and installation tools. Existing guide is reamed oversize so it will accept insert. Insert is then driven into place, trimmed, expanded and finish reamed. Result: better-than-new guide. Photo by Tom Monroe.

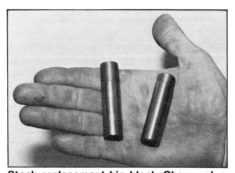

Stock replacement big-block-Chevy valve guides. Guides must be driven out from combustion-chamber side and driven in from rocker-arm side.

Shouldered drift and impact gun is best for driving out old guides and installing new ones.

measure that only delays the inevitable. You don't pay much and you don't get much. The expense comes later when the guides, like knurled pistons, are back to square-one.

To knurl a guide, a special knurling tool is run into the guide. This tool does a job similar to a piston-knurling tool; it raises thread-like ridges on the inside surface of the guide, forcing metal toward the center of the guide bore. Following the knurling tool, the guide is reamed to size, restoring its original ID.

The problem is the same as with knurled pistons. The surface area of the guide is greatly reduced, resulting in accelerated guide wear—even though stem-to-guide clearance was restored. Oil consumption will be just as bad, if not worse than before. So don't knurl the guides if you want a top-notch job. On the other hand, if you are doing a patch job and don't need to get much service out of it, knurl the guides.

Oversize Valves—If, after measuring the valve stems, you find that they are worn and need replacing, you can install valves with oversize stems. Only

the stem is oversize, not the head. To restore stem-to-guide clearance, ream the guides oversize and install the valves.

Chevy offers valves with STD, 0.003, 0.015 and 0.030-in. oversize stems. This method works well and will last just as long as the original valves and guides. Price is the problem as you'll have to buy new valves and pay for the guide reaming. However, if many of the valves are shot anyway, this could be the answer for you. Compare prices of using other methods before making your decision.

Valve-Guide Inserts—At the top of my list of valve-guide fixes is the guide insert. Guide inserts can actually restore guides to better than original. Guide inserts come in two different styles and several materials. The first type is a sleeve that is driven in, the second is threaded in and looks

like a Heli-Coil.

Sleeve-type inserts are available in cast-iron, bronze—sometimes called *phosphor-bronze*—and silicon-bronze. All require the original guide to be reamed oversize so it will accept the insert and so there will be an interference fit between the guide and the insert. After reaming, the insert is driven into place. Some sleeve-inserts are presized, while others need to be reamed or honed after they are installed.

Sleeve-type cast-iron inserts have the same wear characteristics as the original guides and are reasonably priced. Bronze inserts wear better than cast-iron inserts, so it follows that they cost somewhat more. Silicon-bronze inserts have superior wear characteristics; 150,000 miles is normal for a set of these, but they represent another step up in price.

Sunnen makes a kit that uses

Before driving in new guide, guide bore is lubricated to ease installation and seal guide.

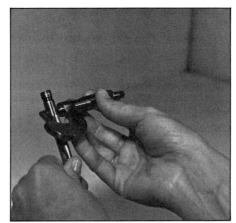

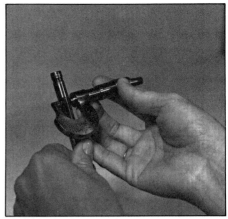

Mike worn and unworn section of valve stem and subtract to determine wear. Highest wear occurs immediately below wear line—measure there. Unworn section is above wear line and below keeper groove. Anything more than 0.001-in. wear means valve replacement.

Using shoulder punch and lead mallet to drive in guide from rocker-arm side. Once guide is installed, valve seat must be ground so it will be concentric with guide bore.

silicon-bronze inserts and requires honing them to size after installation. Some machinists say these are the "cat's meow" because of the smooth finish left on the bronze insert by the hone. The price compares favorably with other reaming and installation procedures and the smooth finish provides maximum thrust area. This translates into long guide and stem life.

Thread-in inserts first require tapping the guides. The tap used must be sharp so that the threads are smooth. If the threads are not smooth, the insert will not mate intimately with the guide, allowing the insert to move up and down with the valve. This

movement pumps oil into the port, so you've got to be careful when installing these inserts. Additionally, a well-cut thread ensures that insert-to-guide contact is good to allow for maximum heat transfer from the valve to the insert, to the guide and, then to the head. If the insert doesn't transfer heat properly, the valve stem will overheat and expand, and you could get a sticky valve.

After threading it in, the insert is locked into place by expanding it with a special tool.

New Guides—The other way to handle the guide situation is to remove the old guides and install new ones. You know what the problem is: The new guide won't center on the original valve seat. However, many machinists don't feel this is a problem.

You need more information before making a decision. Let's look at how new guides are installed.

Big-block guides must be removed and installed in one direction. Driving them out the wrong way will destroy the head. So if you do this job yourself don't blow it and ruin a head while trying to save yourself money.

Always drive the old guides out of the top of the head—valve-cover side—and install the new guides in from the top of the head. Before removing the old guide, measure how far it projects out of the head. Drive the replacement guide in the same *exact* distance.

If you install new guides—or do any guide work—you must grind the valve seats. Otherwise the valve will not even come close to sealing.

Depending on how far off-center the guides are, you'll either have to

grind all day on the seat or they will clean up after a light pass. This is the method used by many shops. When asked what they do about the concentricity problem they just shrug their shoulders and say, "grind the heck out of the seats." It seems to work.

The problem is if the guide is way off center, excessive valve-seat grinding will *sink* the valve and cause problems. First, the stem will project too far out of the top of its guide, resulting in incorrect rocker-arm *geometry*—the area the rocker-arm tip operates on will not be centered on the valve-stem tip. Second, the valve head will move away from the piston. This will increase combustion-chamber volume, resulting in lower compression. Finally, performance suffers due to reduced flow into or out of the combustion chamber because of valve shrouding.

INSPECTING & RECONDITIONING VALVES

Next on your list after valve guides is valve inspection. You may have already declared some valves unusable because of obvious damage; they were burned, warped or badly worn. However, if you haven't given the valves the once-over, do it now.

Inspect the valve-stem tips for damage. If a tip was severely pounded by its rocker arm, it may be *mushroomed*—it'll look similar to the hammer-end of an old chisel. Check the rocker arm if you find a valve tip with this condition.

If you do find a badly worn rocker arm, do not attempt to resurface it. Rocker-arm tips are *casehardened*—a heat-treating process that hardens the surface only. Grinding a rocker-arm

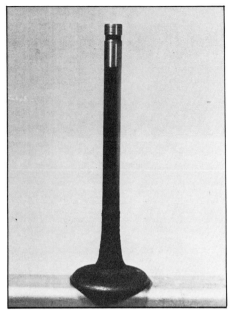

Sometimes there is no need to mike a valve stem to tell it is worn beyond use. This shapely stem was worn more than 0.004 in.

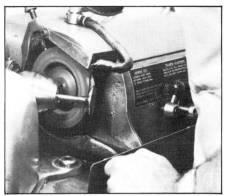

Valve reconditioning. Tip is ground square to stem—just enough to remove signs of wear. After tip is chamfered, face is ground at a 45° angle—with one exception. Truck 366 and 427 exhaust valves are ground at 46°. Photo at right by Tom Monroe.

tip may restore its shape, but the hardened surface will be removed. The rocker arm will wear very rapidly if reused. Lots of scuffing or deep scoring on the valve stem also means a new valve is in order.

Valve-Stem Wear—Although the valve guides and valve stems both wear, the guides wear more. Just the same, let's look at valve-stem wear.

You'll need your 0—1-in. micrometer to compare the worn and unworn portions of the valve-stem. Mike the unworn portion of the stem between the keeper groove and the *wear-line*—where the stem stops at the top of the guide when the valve is fully open. You should be able to feel a small step here with your fingernail. Record your findings. Then measure the valve stem immediately below the wear-line. Subtract this last measurement from the first to get *valve-stem wear*.

Because the greatest valve-stem wear occurs the farthest from the valve head, you should always take the second micrometer reading right below the wear-line in the swept area. The farther away you get from the wear-line, the less accurate your reading will be.

How Much Wear is OK?—If you realize that you have only about 0.002 in. additional stem-to-guide clearance to play with, you can see there isn't much left for valve-stem wear. For instance, if you recondition the valve guides, but run valves with stems

worn 0.001 in., you've already used up about half of the allowable wear. So even though you could use a stem worn as much as 0.002 in. and be within limits, I don't advise it. It wouldn't be long before your engine would start oiling. Use 0.001 in. as your stem-to-guide limit—less if you want maximum durability.

Valve Grinding—For the valves that passed inspection, it's time to recondition them. Reconditioning a valve involves resurfacing its tip and *face*—angle portion of the valve head that closes against the valve seat in the cylinder head. To do this you'll need a machinist with a valve grinder.

The tip end of the valve stem needs your attention first. The tip must be trued first because it pilots the valve in the grinding machine as its face is being ground. An untrue tip will result in an untrue valve face.

As shown in the photo, the valve stem is clamped securely and the tip is passed squarely across the face of a spinning grinding wheel. A very small amount of material is ground away on each pass—only enough to remove small irregularities. Large irregularities, such as mushrooming, require discarding the valve. Finally, the tip is chamfered to remove the sharp edge or any burrs.

Valve faces are reconditioned by grinding away small amounts of material to the correct angle until all flaws are removed. This is also done in the valve-grinding machine. Major components of a valve grinder include: an electric motor to turn the grinding wheel and power the cooling-fluid pump, and a rotating chuck to position and turn the valve. The valve is installed stem-first in the chuck.

The valve is turned slowly while passing over the spinning grinding

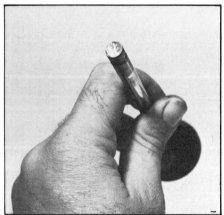

Don't forget to inspect valve-stem tips. A tip worn this badly needs to be refinished.

wheel. While the valve is being ground, cutting oil is pumped over it and the grinding wheel for cooling and to wash away the grinding chips.

It is important that a minimum amount of material is removed from the valve face so there will be sufficient *margin*—thickness of the valve head at its outer edge. If the margin is thinned excessively, the valve will overheat and burn or warp easily, especially exhaust valves. If the margin is thinner than 0.050 in. on an exhaust valve, replace it. Intake valves run much cooler so they can live with a thinner margin—0.040-in. minimum.

All but the truck 366 and 427 exhaust-valve faces should be ground at a 45° angle. Grind the truck exhaust valves at a 46° angle. Later, the passenger-car and light-truck cast-iron-head valve seats will be ground to a 46° angle, giving a 1° mismatch between valve and seat to ensure a gas-tight seal—not guaranteed with identical angles. Seats in aluminum heads are ground to 45-1/2°, a 1/2° mismatch. All truck seats are ground at a 46° angle.

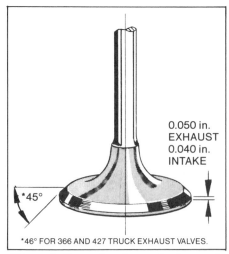

0.050 in.
EXHAUST
0.040 in.
INTAKE

*45°

*46° FOR 366 AND 427 TRUCK EXHAUST VALVES.

After grinding valve face, check margin width. Minimum for intake valves is 0.040 in.; exhaust minimum is 0.050 in. Less margin may result in a burned valve, so replace valve if it is less. Drawing by Tom Monroe.

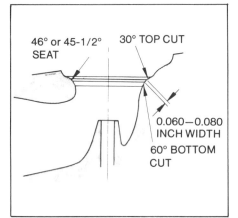

46° or 45-1/2° SEAT

30° TOP CUT

0.060—0.080 INCH WIDTH

60° BOTTOM CUT

Valve seats are ground to 46° or 45-1/2°; 45-1/2° if they are in aluminum heads. Valve-seat width is established by the 30° top and 60° bottom cuts. Outside seat diameter should be about 1/16 in. smaller than the valve face. Wide seats—0.060—0.080 in.—promote long valve life through improved cooling and less valve-face-to-seat pressure. Narrower seats—0.040-0.060 in.—are preferred by racers for maximum sealing and air flow; but long valve and valve-seat life are not the major concern. Drawing by Tom Monroe.

Tools needed to grind valve seats: air drill and mandrel; 30°, 46° and 60° grinding stones; and pilots that install in guides.

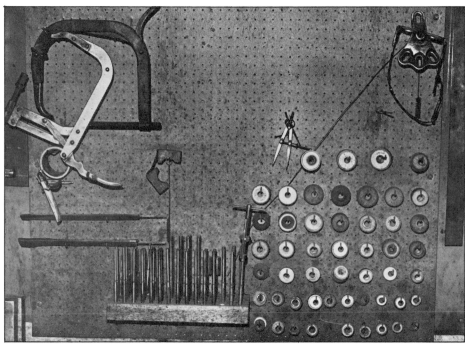

Some equipment needed for grinding valve seats: grinding stones at right, guide pilots at bottom center and dividers above grinding stones. Valve-spring compressors are at top left.

After grinding the valve tips and faces, your valves should be ready to go back to work. Because all major working surfaces on the valves have been restored to like-new condition, you don't have to keep them in order. However, if they are OK and didn't need reconditioning, maintain their order.

VALVE-SEAT RECONDITIONING

Valve seats are ground using special power tools and stones or cutters. If the valve seats have been ground before and there isn't enough seat material left to complete all the necessary cuts, then the old seats are removed and new ones installed. Both grinding and seat replacement are jobs for an expert with the proper tools. The tools are just too expensive and the job requires too much experience for the home mechanic.

Not only does valve-seat grinding restore the seat, it has another very important benefit. Because the grinder *pilots* off a shaft installed in the valve guide, it makes the valve seat concentric with the guide. However, if you didn't do any guide work and the valve seats are OK, you won't have to do any seat work. Any guide work, however, will upset the guide-to-seat relationship; the seats must then be ground.

When correctly done, standard passenger-car valve seats are ground at three different angles, 30°, 46° and 60°. Aluminum-head exhaust-valve seats are ground at a 45-1/2° angle. The valve face contacts the 46° or 45-1/2° cut while the other two angles establish seat width and help smooth the gas flow into and out of the combustion chamber.

To restore the valve seats, the machinist first makes the 46° or 45-1/2° cut with a stone *dressed* at that angle. He then changes to a 30° stone and cuts the top of the seat to establish valve-seat OD. The seat OD is cut about 1/16-in. smaller than the valve face OD. The 30° cut is called the *top cut*.

Next, the *bottom cut* is made with the 60° stone. This greater angle allows the stone to cut below the seat to establish seat ID and width. Seat width should be 0.035—0.060 in. for the intakes and 0.065—0.090 in. for the exhausts.

Special equipment allows the machinist to check his valve-seat work quickly and accurately. He must if he's to stay in business. For instance, the machinist coats the valve-seat area with Dykem-type metal dye—blue or red—so the seat will stand out. This makes it much easier to make measurements because one cut can be distinguished from another. Width and diameter measurements are made with dividers and

Grinding intake-valve seats. Valve-seat concentricity is restored or maintained because grinding stone centers on pilot installed in valve guide. Most machinists grind seats a pair at a time, hence the second pilot in the foreground. After two seats are ground, pilots are moved to the guides in the next combustion chamber.

Lapping valves is done by hand. Abrasive lapping compound is first applied to valve face. The valve is then installed and worked back and forth on its seat with light pressure by hand with a suction-cup-and-stick device. While this procedure shows actual valve-to-seat relationship, it will not correct an incorrectly ground seat, nor will it help a correctly ground one.

If valves or guides didn't need reconditioning or you lapped them, keep the valves in order. You should have indicated FRONT and REAR and which head they're from.

a 6-in. machinist's scale. Additionally, there is a special dial indicator that measures valve-seat concentricity.

Lapping Valves—You may be familiar with *hand-lapping* valves and seats and are wondering where this operation fits into the valve-reconditioning scene. The answer is that hand-lapping is not used in conventional valve work, although racers and other specialists do it regularly.

Lapping valves is done by oscillating a valve on its seat with a stick or valve lapper and suction-cup device. Valve-lapping paste is used between the valve and seat. So, in effect, the valve and seat are ground as a unit.

With the exception of a new valve fitted on an old seat, there's no reason to lap valves. If the valve job is done correctly, the angles will be correct and the valves will seal right from the start. On the other hand if the job is done incorrectly, say the seat and guide were not concentric, lapping will not cure the problem. It will, however, expose the problem. The seat-contact pattern will transfer to the valve face—a real help when grinding. For instance, if this pattern is not consistent on the valve or seat, the guide and seat are not concentric; the seat and/or valve must be reground.

VALVE SPRING INSPECTION & INSTALLATION

Next up for inspection are the valve springs. The following steps must be performed to verify absolutely that the springs are usable. Those that aren't usable *must* be replaced.

Good valve springs are critical to the performance of any engine, particularly the big-block Chevy. Its valve springs must be in top condition because the valve train is relatively heavy, so a lot of *inertia*—resistance to valve movement—is generated, even at low rpm. Weak springs are a problem with the big-block and have caused a lot of performance problems—so give yours the third degree. Be particularly attentive if your big-block is an earlier version.

There certainly is a lot to check when it comes to valve springs. The list includes: *spring rate, squareness, free height, load at installed height, load at open height* and *load at solid height.* If you think a *spring tester* must be used for making some of these checks, you're right. And although it's a lot easier to use a spring tester, some checks can be made with a ruler. From these results, checks requiring a spring tester can be deduced.

SPRING TERMS

To understand what to look for when testing a spring, you should first understand the basic spring terms:

Spring Rate—Basic to the definition of a spring and its function is its *rate.* You can understand spring rate in terms of how strong a particular spring is. A garage-door spring is stronger than a screen-door spring—the garage door has a *higher-rate* spring. Spring rate defines how much *load* a spring applies for a given

amount of *deflection,* expressed in *pounds per inch* of compression for a valve spring. As a spring is compressed, its resisting force increases. Let's look at some real numbers.

Big-block Chevys have been fitted with valve springs of various rates; about 400—650+ pounds per inch (lb/in.). It all depends on the rpm range of the engine. Lower-output passenger-car big-blocks are not designed to turn much over 5000 rpm, while the special, super-heavy-duty engines are designed to turn as much as 7000 rpm. Obviously, it is going to take relatively high-rate springs to control the valves of the higher-rpm engine. If lower-rate springs were used the valves would *float* at high rpm.

Valve float occurs when the inertia load of the valve train exceeds the ability of the valve springs to close the valves—the springs cease to hold the lifters against their cam lobes. When this occurs, compression and combustion pressures are lost out the intake or exhaust ports, or worse, the pistons contact the valves. In the first instance, power is lost; in the second, valve-train or piston damage results.

Engineers try for the lowest-possible spring rates, even on race engines. Only enough "spring" to close the valves, not much more. This is because it takes horsepower to turn the camshaft and raise the lifters and valves against their springs. Too-stiff springs mean wasted horsepower and unnecessary valve-train wear.

Valve-spring squareness is easily checked with a carpenter's square. Rotate spring until you find maximum lean. Distance from vertical leg of square to top coil should not exceed 1/16 in. If it does, replace spring.

VALVE-SPRING SPECIFICATIONS

Engine	Year/HP	Height (in.)	INSTALLED Load (lb)	Min. Load (lb)
366/ 427T	all	1.80	90	85
396	65-69/ 265, 325	1.88	84-96	80
	65-69/ 360,375, 350,425	1.88	94-106	89
402	70-72	1.88	69-81	66
427	66, 67/ 385	1.88	94-106	89
	390 400 435	1.88	94-106	89
	425	1.78	37-45	35
		1.88	69-81	66
	68-69/ 385, 390 400, 425 435(3x2Bbl)	1.88	94-106	89
454	70/ 345	1.88	69-8l	66
	70/ 360	1.78	26-34	25(inner spring)
	390 450	1.88	69-81	66
	70/	1.78	37-45	35(inner spring)
	460	1.88	69-81	66
	71	1.78	26-34	25(inner spring)
		1.88	69-81	66
	73, 79	1.88	74-86	70
	80	1.80	84-96	80

What this means to you is there is little reserve strength in the valve springs. After repeated hot/cold cycling and millions of openings and closings, springs *fatigue,* or lose their load-producing capability. More exactly, the springs will *sag*—much like a chassis spring. When this happens, the valves will float at lower and lower rpm.

Free Height—Free height is simply the unloaded height, or *length,* of a spring; how high, or *tall,* it is when sitting on the bench. Remembering that the force or load exerted by a spring is directly proportional to how much it is compressed, you'll understand why a fatigued spring is shorter—has less free height—than it was when new.

Because you don't have a spring tester to determine a spring's rate—or load at different heights—use free height to judge valve-spring condition.

Load at Installed Height—Load at installed height is the load a valve spring exerts when installed in the head and the valve is closed. It is also a specification used for checking valve springs. An average figure for big Chevys is 95 lb at 1.88 in. of compressed height.

Load at installed height is important because, like free height, it gives a good indication of a spring's condition. For normal service the minimum load is 5% below the spring rating, or 90 lb in our example. If you want your engine to be capable of revving to its design limit, stick to specified load at installed height, not the minimum limit. This holds true for all performance big-blocks—L-34, L-78, L-88, LS-5, LS-6 and up—regardless of intended use.

Load at Open Height—Similar to load at installed height, load at open height occurs when the valve is fully open—or when the valve spring is compressed more than its installed height by the amount of maximum valve opening.

Typically, load at open height is 300 lb at 1.38 in.—or 205 additional pounds for 0.50-in. additional compression. Again, if your big-block won't be used in high-performance applications, this figure can be 5% lower than rated specifications.

Don't waste your money on new springs if your car will only be used for going to the store and back and you find a couple of "5% springs." On the other hand, don't think high rpm means racing—if you use the lower gears for passing or hill climbing, then your engine needs springs that are right on spec. Racers usually want valve springs within 10 lb of each other—the 5% allowance is 15 lb for a 300-lb spring. And, if you're building a high-performance engine, check new springs on a spring tester—don't take anyone's word for it.

Solid Height—If a valve spring is compressed until each of its adjacent coils touch and it cannot be compressed any farther, the spring is at its *solid height*—it is *coil bound.*

In the real world, a valve spring that *goes solid,* or *coil binds,* before its valve opens fully creates great havoc with its valve train. Usually it's the pushrod that bends. A not-too-uncommon occurrence, coil binding is caused by a cam with more lift than the original cam, or by a spring *shimmed* excessively to bring it within specifications, page 104.

Squareness—How straight a valve spring stands on a flat surface is called *squareness*. If a spring *leans* too much—1/16 in. or more measured from the top coil to a vertical surface—it should be replaced.

SPRING TESTS

Now that you've been bombarded with the basic valve-spring terms, apply this information to put your valve springs through several tests. Check for: squareness, free height, installed load, open load and solid height.

Unless you are among the fortunate few, you probably don't have a spring tester. The industrial type pictured will chop a $1000 bill right in half—this puts it out of the occasional-user league.

Fortunately there is a valve-spring tester that is within the average do-it-yourselfer's range. Although not as convenient to use, all you'll need, in addition to the tester, is a vise and a vernier caliper. Clamp the spring and tester in the vise, compress the spring to its installed or opened height and read spring load directly, with the calipers.

If your local auto-parts dealer or speed shop doesn't sell such a tester, one's available from: Goodson, 4500 West 6th Street, Winona, MN 55987; or C-2 Sales & Service, Box 70, Selma, OR 97538.

Squareness of a spring is the easiest check, so let's start there. The equipment needed is a carpenter's framing square—or any right-angle-measuring device will work. Stand the spring upright in the corner of the square. Turn the spring until you find maximum *lean*—distance between the top coil and the vertical leg of the square. Maximum lean is 1/16 in. Replace the spring if it leans too much.

Free height is a fairly accurate indication of a valve spring's condition. Although you won't be able to measure spring loads, you can measure *free height* with a 6-in. machinist's rule. If free height is within spec, chances are good that installed and open loads will be OK.

Find the specifications for your springs in the accompanying table, then measure free height. Chances are springs that are short—have *sagged*—by more than 0.100-in. of

Checking spring load. If load is not enough, shims are added. After shimming, spring is checked at open height for *coil bind*—coils touch and spring cannot be compressed any farther. There must be at least 0.012 in. between coils at open height.

their free height, have fatigued. They should be replaced or checked in a spring tester for installed and open loads.

A good rule of thumb is to replace *any* spring that is short 1/8 in. or more. If the spring is only 1/16-in. short it may be used in an engine destined for light service. It would be foolish to install such a spring in an engine that will be operated in the upper-rpm range.

You may have weeded out a spring or two and replaced them. Just because they are new doesn't mean you shouldn't check them. Put them in your lineup and perform the remaining tests on them. That brings up another point: spring order. You should've kept the springs in order, unless you reground the valve seats or valves. If they got mixed up during cleaning or whatever put them back in line and maintain their order from now on. Spring inspection from this point on requires a spring tester. If you don't have one, gather up your springs and take them to your engine machinist.

Springs that fail the free-height test by a 1/16 in. or less can be *shimmed* back to original height. Shims are available at local parts houses for adjusting spring height. VSI lists three shims for the big-block: 0.015, 0.030

Checking installed spring height with a 6-in. rule. You must use valve for that guide and 0.015-in. hardened-steel shim found under all springs, or rotator for 366 or 427 truck. Keep springs in order from this point on. With the keepers and retainer installed, pull up firmly on the retainer and measure from top of the shim or rotator to underside of retainer. Refer to spring chart for specified installed height and add shims as necessary.

and 0.060 in. You must have a spring tester to see if a spring will work when shimmed, and how thick a shim it needs.

If you shim a spring to restore its installed and open loads, check for coil bind! Because the spring will be compressed more than its specified open height, its reserve travel is reduced when the valve is open—distance between coils will be reduced. This is especially true of high-performance engines fitted with high-lift camshafts. There must be 0.012 in. between the coils when the spring is compressed to its open position.

There are a couple of reasons why it is better to replace a spring than to shim it. First, a new spring has its entire life ahead of it, the old spring is already on its way out—that's why it needs shimming. Second, a spring shimmed to increase its loads is compressed more. Thus the constantly greater compression subjects the spring to greater stress than it was designed for. This fatigues the spring further, and at an increasing rate.

Unless you are patching a big-block for a few thousand more miles—knurled valve guides, pistons and such—replace valve springs before you shim them to obtain closed and opened loads. If you do shim a spring, wire the shim to the

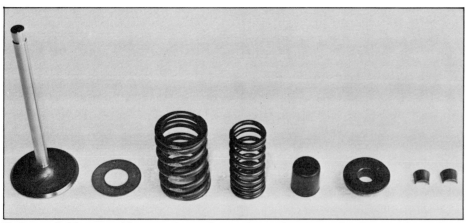

Parts needed to assemble one valve: valve, spring seat, outer spring, inner spring, valve-stem seal, retainer and keepers. Make sure seals fit inside valve springs. There are seals for single springs and others for double springs.

Valve installation starts with spring seat. If a shim is needed, it goes on next.

Slip valve into its guide, then oil the stem. This will allow easy installation of the valve-stem seal.

Hold valve against its seat and carefully push seal over valve stem. If required, use cellopane tape over stem tip and keeper grooves to prevent the seal from being cut as it is passes over keeper grooves.

spring so they don't become separated. Otherwise you'll have to re-check the springs.

CYLINDER-HEAD ASSEMBLY

To assemble the cylinder heads, you'll need a valve-spring compressor, a ruler and *possibly* an assortment of shims. Don't buy the shims until they are needed—chances are they won't be.

Installed Height—Let's shift our attention to spring *installed height*—height of an installed spring with its valve closed. Directly related to load at installed height, installed height is important to ensure that the springs exert the proper closing force to prevent valve float.

Don't be confused about the possibility of having to shim the valve springs again. The shimming you may have done was to make up for sag or to restore installed and open loads. This checking and possible shimming is to ensure that the distance between the spring seat and its retainer is at the specified installed spring height.

For instance, if a valve stem extends too far out of its guide—if the valve seat or valve face needed considerable grinding—the spring will not be compressed to its installed height. The spring must be shimmed to compensate for the difference in installed height so installed and opened loads will be obtained. Therefore, you may need to add more shims to springs that already have them; one set for installed or open load, and possibly one for installed height.

Check Installed Height—Installed height is checked with the valve installed in *its* guide without the spring. Start by placing the hardened-steel shim—rotator in case of a 366 or 427 truck—on the spring-seat pad. Insert the proper valve into the guide. Lightly oil the valve stem and install the valve seal using the plastic sleeve-installer, if provided. If not, use a strip of Scotch-type tape.

You don't have to remove the valve later, so installing the seal is one assembly operation that's finished. In addition, the seal helps to hold the valve in place.

Now install the retainer with its keepers. Pull the retainer up snugly against the keepers. While lifting firmly on the retainer, measure the distance from the spring-seat shim or rotator to the underside of the retainer. This check must be accurate, so make sure the lighting is good. One of two methods can be used for measuring installed height.

If you have a telescoping gage, use it and a 1—2-in. micrometer. If not, a small rule or the more accurate vernier calipers will do. Record your measurements for each valve.

Second, an easy way to measure installed height is with a gage made from a length of heavy wire or welding rod. Cut a piece of wire with your wire cutters, then file it to the exact specified installed height—use 1—2-in. micrometers for checking gage length.

Check installed height by standing the gage between the retainer and the shim or rotator. Use a feeler gage between the gage and the spring retainer to determine needed shim thickness.

Subtract specified installed height from your measurement to find the shim needed—if one is. Ignore anything less than 0.015 in. Compensate for anything over 0.015 in. with a shim of the same or *less* thickness.

Because I told you not to buy any shims until you knew what you needed, I can assume that you don't have any. So record the thickness needed on a narrow piece of paper and slip it under the valve spring that needs the shim. Also, write it down

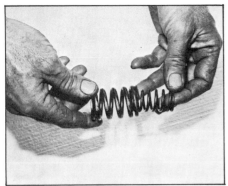

If your engine uses double springs, don't forget the inner spring. Just slip it inside outer spring.

Fit spring/s over valve stem and place retainer against spring/s. Head rolled over on its exhaust ports is the best position for head assembly from this point on.

Compress springs and work keepers into their grooves. A dab of grease on keepers will keep them in place. Slowly release compressor and check that the keepers remain seated in their grooves.

on your shopping list. Later, when you come back from the parts house, it will be easy to match the shim with its spring.

Retainer-to-Guide Clearance—Next you need to make sure that the bottom of the retainer clears the top of the valve guide and its oil seal by at least 0.060 in. Do this by opening the valve—while holding the retainer and keepers in place—to its full-open position. Use a machinist's 6-in. rule against the retainer to check valve opening. Eyeball the retainer-to-guide clearance. If it appears to be remotely close, measure it with a feeler gage.

If there's insufficient retainer-to-guide clearance, the top of the guide must be machined shorter. Have a machine shop do this. Even if you aren't changing to a higher-lift cam, you could get into trouble here. There are different-length guides available for the big Chevy. Consequently, if the wrong guides were installed, you could have a real problem. Admittedly this is a long shot, but it happens, and checking is a relatively simple operation.

Cylinder-Head Assembly—After purchasing the shims, you should have everything needed to assemble the heads. If you have everything laid out, head assembly will go quickly and with little drama. Don't worry about more than one shim under some of the springs. You could have shims for free height, installed height or no shims at all, except for the one that acts as a wear surface between the head and spring seat.

Organize all parts in front of you. Have the heads with the valves and shims or rotators already installed, and their springs, and any additional required shims in order. Remember,

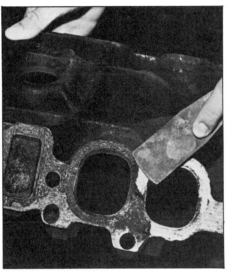

To restore the intake manifold, about all it needs is a thorough cleaning. If you are going to hot-tank it, remove all exterior hardware that could be harmed by the hot-tank solution. However, unless the manifold is really dirty or it is aluminum, scraping the gaskets and solvent cleaning will do the job.

springs, shims and valves must be installed exactly as they were checked. Each must be installed with its valve. You'll also need the valve-spring compressor and an oil can.

Remember to compress a valve spring only enough to install the keepers, no more. After the spring and retainer are compressed, install the keepers. A dab of grease on the keeper groove will hold the keepers in place so you can use both hands to release the compressor.

After they are assembled, store your cylinder heads in a dry, dust-free area. Protect the machined surfaces and valve-train parts with a light coat of spray-on oil and bag 'em.

INTAKE MANIFOLD
Because of its close relationship

Assembled head is ready for installation. For now, store heads in plastic bags so they will stay clean, dry and rust free.

with the cylinder heads, clean up the manifold so it will be ready for installation after the heads are installed.

Ah, the beauty of non-moving parts! There isn't much to do to restore the intake manifold to like-new condition. If it is cast iron, the quickest and easiest way to clean it is in the hot-tank; remove the divorced choke and any aluminum fittings first. Clean an aluminum manifold by brushing it with solvent. Use a wire brush to remove the baked-on grime.

Water passages could use the attention of a blunt screwdriver and rat-tail file to remove heavy deposits. Besides inspecting the runners for loose objects, there isn't any cleaning to do there. The important areas are the two gasket surfaces along each side and the carburetor-mounting surface.

If you hot-tanked the manifold, the divorced-choke spring needs to be refitted to the manifold. Make sure the spring is still intact before tightening the mounting screw. There are many other small fittings to be installed on the manifold, depending on engine model. If it was on the manifold before, reinstall it now.

7 Engine Assembly

1969 was the last year the three 2-barrel 427 was available, both in 400- and 435-HP versions. A big-block Corvette with this induction system is highly desirable in the collector market. Photo courtesy of Chevrolet.

From here it's all downhill. All that's left to do is bolt your engine back together and install it in the chassis. By now most of the running around is done and the parts are clean. It's time to put your engine together.

THINGS TO HAVE

Sealer—Although I said most of the parts chasing has been done, there are several items you'll need by your side during the engine assembly. Buy them before you start. First the sealants: Nearly every gasket in your engine requires some kind of sealant to do its job. You can't use the same gasket sealer for all jobs, so it isn't as simple as using up that old tube of "stick-em" in the junk drawer.

One of the best is RTV (Room Temperature Vulcanizing) silicone sealer. RTV is best for sealing liquids, but it doesn't hold gaskets very well when you're trying to put two parts together. An adhesive gasket sealer works better when you're trying to hold troublesome gaskets in place. A good adhesive-type sealer is 3M Super Weather-Strip Adhesive, part number 08001. This one really is a weather-strip adhesive—although it works fine for gaskets. Or, use GM Gasket Sealing Compound, part 1050805, which comes in a 15-oz aerosol can.

You'll find that 3M's Weather-Strip Adhesive is easier to use, but it will sure make a mess if you ever have to take the engine apart. You could use OMC's (Outboard Marine Corporation) excellent Adhesive Type M, or Ford Gasket and Seal Con-

tact Adhesive, part D7AZ-19B508-A. Two other readily available sealers are Permatex High Tack and Gasgacinch.

Oil & Grease—In addition to all the sticky stuff, you'll need lubricants. First on the list is a quart or more of the engine oil you'll be using in your new engine. There are many different brands of motor oil on the market and some people are partial to one particular brand. The important part is that your oil should carry at least the *SE designation*. If you use an SF oil, that's even better, but nothing less than SE should ever go in your engine.

Under no circumstances should you break in your engine on a non-detergent oil, and then switch to a modern, "graded" oil. The rules of 30—40 years ago don't hold true; today's rings, cylinder walls and other internal engine parts break-in right away. There's no need for some trick engine oil for break-in. Use SE or SF

Lubricants needed for assembling your engine. Anti-seize compound is especially useful for the head bolts because it also acts as a sealer. Head bolts thread into water jacket and must be sealed. Moly-disulfide (MoS$_2$) grease is essential for cam and lifter break-in.

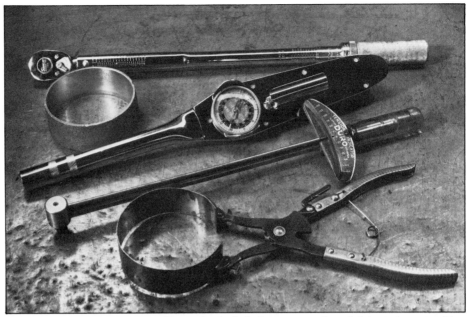

Special tools needed to assemble engine include torque wrench and ring compressor. Pictured here are three torque wrenches. From top to bottom they are: clicker, clicker with dial and light, and beam type. Cylinder at upper left is a cone-type ring compressor for installing pistons. More common—and universal—plier- type compressor is at bottom.

oil for break-in and forever after.

In addition to motor oil, get some EOS (Engine Oil Supplement) from your Chevy parts man. Ask for EOS, part 1051396, in the 16-oz pull-top can. This is a fantastic oil additive that you might want to use long after your big-block is back in service. Another "must-have" is a supply of molybdenum-disulfide grease, or *moly grease* or *MoS$_2$* as it is often called. Poke around the parts house until you find a small tube of this cam-saving grease.

Finally, fill your squirt can with fresh motor oil of the same type you'll use in the engine.

This list of lubricants may seem long and possibly redundant, but there's nothing foolish about paying strict attention to all the lubricants in your engine for start-up. These first minutes of engine operation are very critical. What takes place inside in this short time can determine to a great extent just how long your new engine is going to last.

Most important of your start-up concerns is cam/lifter break-in. It is so easy to wipe-out a cam lobe by improper lubrication when the engine is first run. It takes less than *a minute* to destroy a lifter and cam lobe—that is no exaggeration! The point is, *don't skimp* on engine lubes.

Tools—Here's the list of specialized tools you'll need for engine assembly. Start rounding them up if you don't have them at hand.

Torque Wrench—First on the list is a torque wrench. I've mentioned this tool before, so I won't go into a full description again. You must realize that there is no way you are going to put an engine together without one. I don't care how experienced or talented a mechanic may be, there is no human way to "feel" the correct amount of force necessary to torque a nut or bolt to the *exact specification* required. You may come close but you'll never hit it right on. And that's what you must do when building an engine; get bolt torques right on specs.

Ring Compressor—To install a piston, the rings must be squeezed together so they will go into the bore. You will need a ring compressor for this chore. Cone, plier and cylinder-type ring compressors all work well. I doubt if you'll find a cone-type compressor—only machine shops have this style—but your chances of getting a plier or cylinder-type are about equal. The plier type squeezes a sheet-metal C around the rings. The cylinder type wraps completely around the piston and tightens with a series of bands. Pro mechanics seem to prefer the plier-type because it won't bind-up like the cylinder type; and it's quicker. Use what you can get.

Plastigage—A smart mechanic doesn't believe everything he reads on parts boxes—including the size of bearings. So many mechanics have

been fooled by the wrong bearings in the right boxes it would take a book this size to list their names. To stay off that list, buy a strip of *Plastigage*. This is a length of precision thickness wax, believe it or not, and you'll use it to check bearing clearances. Purchase the green variety, 0.001—0.003-in. You don't need much, one strip is plenty.

Make sure the Plastigage you purchase is fresh to ensure the readings you get will be accurate. Dried up, hard Plastigage will show more than actual clearance.

Ring Expander—A tool that isn't mandatory, but one that sure makes life a lot easier and potentially much less frustrating, is a ring expander. A specialized pair of pliers, the ring expander holds and spreads—expands—a piston ring so it can be slipped over a piston and into its groove.

If you didn't use a ring expander for removing the old rings, think back to how your thumbs felt when removing them. And the new rings are even stronger. Attempting to install 16 compression rings—24 in a truck—will wear two painful holes in your thumbs you won't soon forget. In addition to the physical pain, think about the mental pain of having to lay out more bucks to replace a broken ring. It is all too easy to break a ring

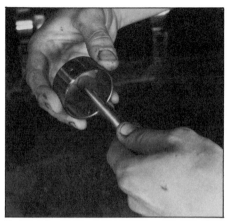

Deburr cam bearings before installing them. If you don't, cam installation could be a real problem. Hold bearing scraper or penknife at a 45° angle to bearing. One pass around each end of bearing should be all that's needed.

Early engines—1965–'66—use grooved rear cam-bearing journal and bearing. At right are the '67 and later counterparts. Ungrooved cam and bearing cannot be used in '65 and '66 big-blocks; grooved cam can be used in later big-blocks.

Using drive-bar and mandrel to install cam bearings. It is very important to install cam bearings so oil holes in the bearings and block align. If they don't, nothing above the crankshaft will receive lubrication.

when attempting to install it by hand. And, you don't want to have to buy another complete set just for one ring, and spend extra time chasing it down. Beg, borrow, buy or rent a ring expander for installing your rings.

Cam-Bearing Installer—Finally, if you still haven't had your cam bearings installed, get your block and bearings to the machine shop and have them driven in. If you still want to try this yourself, you'll need another special tool—a cam-bearing installation kit.

Soap—Finally, your hands must be clean for assembling the engine. Have some abrasive soap handy so you can keep your hands absolutely clean. Otherwise, dirt on your hands will end up being circulated throughout your engine's lubrication system.

CAM-BEARING INSTALLATION

Cam-bearing installation starts with assembling the tools. There are various types of cam-bearing installation tools, but the two common ones are: threaded-puller, drive-bar and mandrel types. No matter which style you use, three steps are important. First, the bearings must be driven in squarely; second, the bearing oil holes must line up with those in the block; and third, the bearings must not be damaged during installation.

Size Difference—Although the box the bearings come in should list the bearing-identification numbers and give their locations, you should know about the differences between them.

All cam bearings are the same diameter, but their lengths and oil-holes are different. Cam-bearing 1 is

short by almost 1/8-in. On 1965—66 models, bearing 5 has several oil holes and a groove. On '67-and-later models, bearing 5 has a single 0.116-in. oil hole and no groove. Bearings 3 and 4 are identical, while bearing 2 has a chamfered oil hole.

Chamfer Bearings—Before you start hammering the bearings home, clean up their edges. This is called *chamfering* the bearings, because that's what you do to them, put a chamfer on their inner edge. As the bearings come out of the box they probably have several burrs around their inner edges. If you leave these burrs in place they can hang up the cam when you try to install it.

The time to cut the chamfer is now, not after they are installed. Chamfer the bearings with a pen knife or, if you have a really complete tool box, use your *bearing scraper*. A bearing scraper is a a triangular section of steel with sharp edges. It looks a lot like hollow-ground three-sided file without teeth.

Hold the bearing and scraper 45° to each other while running completely around the bearing inner edge. Don't remove much material! Just enough to remove any burrs—and that isn't more than a couple of thousandths. Think of the thickness of a 0.002-in. feeler and you'll see that I'm not talking about removing much metal. When you run a fingernail around the finished edge, it shouldn't hang up.

Bearing Positioning—Roll the engine block upside down so you can look into and reach through the crankcase opening during cam-bearing installation. You must guide the bear-

ings while driving them in, as well as sight through the bearing bores to make sure they line up.

Set the bearings on the bench in order of installation so you won't have to fish around for them later. Size differences are obvious, but if you're uncertain, check the bearing shell and the box. The box should list the bearing diameters and their positions so you can be positive.

Pay attention to the oil holes in the cam bearings. Bearing oil holes *must* line up with those in the engine block or you'll have big troubles. If an oil hole is covered, that cam journal will not get *any* oil, resulting in complete destruction of the bearing and probably the cam journal as well.

Cam bearings have small holes and the block has larger ones, so you can get a bearing insert turned a little, but not much. Therefore you should be extra careful.

Grooved Journal—Although I mentioned the rear cam-bearing differences earlier, I've saved the full discussion until now. 1965 and '66 engines oiled the lifters via a grooved rear camshaft journal and bearing. 1967 and later blocks have an annulus groove behind the rear cam bearing that feeds the lifters. On these later engines, the rear cam journal and bearing are exactly like the front four.

There is no way to use a later cam in a '65 or '66 block without machining a groove in the rear journal, so that is simple enough. If you did fit the later, non-grooved cam to an early block, there would be no lifter oiling.

Early, grooved camshafts can be used in later blocks. No modifications

Drive in cam plug with a large-diameter driver so plug will not distort and leak. Correctly installed plug will be flush with rear face of block.

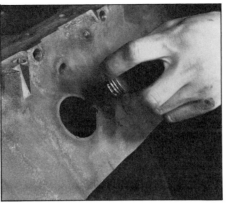
Seal core-plug bores. This will help installation and help ensure that the core plugs will be leak-free.

Special tool my engine machinist made for installing core plugs. You can do the same job with a large-diameter punch or an extension and large socket. Don't use a small punch in the center of the plug. It will distort plug, causing it to leak.

are necessary to the camshaft, but the rear cam bearing must be the later, non-grooved type. Before the rear cam bearing is installed, the oil hole must be soldered shut, redrilled to 0.060-in.—1/16-in. is close enough—and then installed in the block. The original oil hole is too large and will allow a large internal oil leak that will disrupt engine oiling.

Bearing Installation—Now you're all set to install the bearings. Start from either the front or rear of the block and work toward the center; then turn the block around and work from the other direction. This keeps the installation tool passing through as many bearing bores as possible. It gives you the maximum number of bearing bores to sight through—reducing the chance of getting a bearing cocked.

For example, you could start with the rear bearing, 5. Then install 4 and 3, the center bearings. Turn the block around and install 1, and then finish with 2. This is especially important when using an installer that lines up by eye. You'll have to be careful not to damage some of the bearings as you work through them.

Select the bearing and slip it over the installation-tool mandrel. If your mandrel is the expanding type, expand the mandrel until it supports the bearing ID. If you're using solid mandrels, use the one that fits closest to the bearing ID. If the mandrel is too loose, wrap it with masking tape to take up excess clearance.

Position the mandrel-and-bearing combination next to the bearing bore. Line up the bearing and block oil

holes so when the bearing goes in place they will match. With drive-type installers the bearing should be on the same side you want to drive from, the pull-through threaded type start from the opposite side, the one you'll be pulling from.

Start the bearing into its bore *without lubricant*. The bearing back and bore must be dry and clean. Stop just as the bearing starts in its bore and verify that it is going in straight. Straighten the bearing if it's crooked by compensating with the angle of the mandrel. The sooner you check the bearing on its way in, the easier it will be to correct any deviation, so don't be overanxious to slam the bearing in.

Continue to drive in the bearing until it looks centered. Check the oil hole for any *shrouding*. This is easily done with a flashlight or droplight. Shine the light up from the crank journal into the cam-bearing oil passage while sighting into the oil hole from above. Also check from above with the light. You should see no overlapping between bearing and block.

Cam Plug—After you are satisfied that all the bearings are located correctly, install the *cam plug*—the large cup plug at the rear of the block.

Wipe off the plug and apply a sealant to its outer edge. RTV silicone sealer works well, or apply another non-hardening sealant, such as Permatex No. 2. It is important to seal this plug or you'll have an oil leak inside the bellhousing. If the engine will be mated to a manual transmission, oil will make the clutch slip and grab. At the very least you'll have an oil leak .

Drive in the plug with a socket or a

flat disc. The socket should fit just inside the plug flange, the disc should cover the plug. The disc works well because it will set the plug to its proper depth—flange flush with the block. The easiest method is to start the plug with the socket, then finish with the disc. You can use almost anything for the disc. It should be about 3-in. in diameter, and flat and thick enough so it won't bend under hammer blows.

Core Plugs—There are eight additional plugs in the block that you might as well install now while you have some sealant close at hand. These are the core, or water-jacket plugs. Two in the rear of the block flank the cam bearing, three are in each side of the block. Seal these with RTV or non-hardening sealer and drive them in with your large punch or socket. Drive these plugs in so their flanges are just past the chamfer.

Oil-Gallery Plugs—If you removed the oil-gallery plugs—the Allen and square-head threaded plugs around the block—coat them with non-hardening sealer, starting on the second thread, and run them in quite snugly. Don't put sealer in the threaded hole. The sealer will end up in the oil passage. For the same reason, don't use Teflon tape to seal oil-gallery plugs. It is too easy for a piece of the tape to end up in an oil gallery. Little pieces of sealer or Teflon tape floating around in your engine oiling system can do a lot of damage by obstructing oil flow.

Oil-gallery plugs have pipe threads designed to seal when tightened, so you'll have to apply some muscle on

VENTED OIL-GALLERY PLUGS

Some later big-blocks incorporate an oiling modification you might want to make, particularly if yours has a gear-drive cam. Oil-gallery plugs that install adjacent to the front cam bearing have a 0.030-in. hole drilled in their centers. These *bleed holes* were put there for several *rumored* reasons. The most common rumors are to bleed air from the lifter galleries, lubricate the timing chain or lubricate the cam timing-gear thrust face.

The last reason—to lubricate the gear thrust face—is the most plausible. The sprocket of a chain-drive-cam is lightly loaded against the front of the block by thrust load from the oil-pump/distributor gear. However, wear as a result of thrust load from the factory-installed helical gear would be much greater without additional lubrication from the drilled oil-gallery plugs.

If your engine has a gear-drive cam or you just want to provide additional lubrication for the chain and sprockets and your plugs are not drilled, drill them yourself with a 1/32-in. drill. That's a very small drill, so don't bear down too hard and break it.

Do not drill the plugs while they are installed, otherwise metal chips will end up in the oiling system. For the same reason, deburr the plugs after you drill them.

If you do have the drilled plugs, be sure they are reinstalled in the front of the two lifter oil-gallery holes or you will have a massive oil leak where you don't want it.

Oil-filter adapter is secured to block with two bolts.

Coat cam lobes and fuel-pump eccentric evenly with moly before installing cam. Lube bearing journals with oil. Also oil cam bearings.

the last turn or so. Don't skip the Allen-head plugs. Make sure each plug is tight before you install the next one.

There are nine Allen plugs and one square-head plug. If you have any plugs left over in your parts box, check the timing-chain area next to the front cam bearing. Two Allen-head plugs go there. Read the sidebar about drilled plugs that go here. Just outside the timing-cover area at the lower left—on the right as viewed from the front—one Allen-head plug fits into a recess.

Spaced along the bottom left side of the block at each main-bearing web are threaded holes for four more Allen plugs. The last two Allen plugs install next to the cam plug at the rear. A little lower and to the left is the square-head plug. These are all the oil-gallery plugs on most big-blocks.

Some high-performance blocks were drilled and tapped for oil-cooler lines right above the oil filter. Don't forget to plug these holes if you removed the plugs or have no cooler.

Oil-Filter Adapter—If you removed the oil-filter adapter, reinstall it. Use the two 1/4-20 bolts, but be careful not to overtighten them. The aluminum adapter is fragile and will crack easily if you cinch it down too much.

INSTALL CAMSHAFT

Camshaft installation and preparation are critical to camshaft longevity for the first minutes of engine operation. Coating the cam with oil and slapping it in the block is good for about 30 minutes of cam life. I'll wager you're interested in a little longer cam and lifter life for your time and money.

Big-block Chevys use strong valve springs that really load the lifters against the cam lobes. You've got to protect the lifters and lobes from excessive wear with high-pressure lubricant during initial engine startup.

Moly is the answer to high-pressures and it's the only thing you should use to lube cam lobes. You'll be protecting the cam and lifters along with the rest of the engine. Moly protects the entire engine by keeping cam and lifter wear to a minimum. If the cam and lifters wear excessively, the minute, broken-off particles will be carried throughout the engine by the oil. The filter will get only so many of these particles, leaving the rest free to get in between close-tolerance parts or embed in the bearings. Virtually all of these particles will be drawn through the oil pump.

Lube Camshaft—Before giving the cam its coating of moly grease, wipe it down with solvent or lacquer thinner using paper towels. Unlike the typical shop rag, lint from a paper towel is oil-soluble. Lint from a cloth rag will cling to everything inside the engine and will help clog the oil filter.

Now that you've got the cam clean, make sure your hands are clean. Scrub them with an abrasive soap to remove as much dirt as possible. If you don't, all the clean moly and oil

Feed cam into block like you removed it—very carefully. Don't bump bearings with the cam lobes. This job is a little easier if you're assembling engine on a bench. Position engine on its rear face and lower cam in from the top—it's easier to control.

you'll be lubricating the cam with will act as a solvent, removing the dirt from your hands and depositing it inside your engine! Besides, from here on you shouldn't get very dirty.

Coat the cam lobes with moly. You can either use a short, stiff-bristled brush or your fingers. Go completely around each lobe, coating its entire surface with moly grease. Don't gob the moly grease on, but apply an even, medium coat—don't forget the fuel-pump eccentric.

Next, give the cam journals and

Always check bearing sizes and numbers against those on the box. Getting the wrong bearings in the right box is a common problem, so make sure you get what you asked for *before you leave the counter.* This main bearing is STD (arrow).

bearings in the block a liberal coating of oil from your squirt can. Spread out the oil with a fingertip.

Install Camshaft—Position the block so you can easily feed the cam into it. Just as with removal, the easiest position is with the block on its rear face, timing end up. This allows you to lower the cam through its bearings while guiding it with your free hand. Second-best position is with the block rotated upside down on the engine stand. Make sure you can get your free hand in through the crankcase to guide in the cam.

Run a long bolt in the nose of the cam to serve as a handle for more control. Holding the cam by the bolt at one end, feed it into the block carefully. Note: On a gear-driven cam, you'll have the gear to use as a handle. Avoid contact between the cam lobes and the edges of the cam bearings. The cam lobes will damage the bearings if you are careless. With your free hand inside the engine, guide the rear end of the cam.

When the rear journal is ready to mate with its bearing, the other four journals are also ready to slide into their bearings. If you chamfered the bearings, this is where you get your reward. The cam should slide easily into place. There may be a little hang-up, but a slight push should easily overcome it.

With a gear-drive cam, install the thrust-plate bolts and torque them 7 ft-lb, or 80 in-lb.

Caution—Now that your well-lubed camshaft is in place, exercise care as you handle the block. Big-block Chevys do not have a thrust plate—unless the cam is gear driven—so there is no easy way to hold the cam in without the timing chain and sprockets. As you move the block to install the crank, pistons and other internal hardware, make sure you don't turn the front of the block down or the cam will slide out. When you have your hands or a tool inside the crankcase, be careful not to bump the cam. If you do, make sure it didn't move out of position. This could save you from running a connecting rod into the cam later on.

CRANKSHAFT INSTALLATION

With the camshaft safely home, it is time to "lay the crank." You'll need the main bearings, rear crank seal plus the main caps and bolts. Many times the machine shop that did your crank work will want to sell you the bearings along with the crank work. This is fine as long as the bearings are the right size, a name brand and not *fully grooved.* Expect to get the bearings with a crank kit.

Before getting into bearings, let's first look at the basics. Chevrolet bearings are excellent, of course. Those from Federal-Mogul are also well respected. There are many other good bearings, but these are two of the best. Chevy does not offer fully grooved bearings, so there's nothing to worry about there. Federal-Mogul does, part number 4919M. You do not want fully grooved bearings!

Grooved Bearings—Fully grooved bearings have an oil groove in *both the upper and lower bearing halves.* Stock bearings have grooved upper inserts, but the bottom halves—the ones that go in the bearing caps—must never be grooved. If yours are, return them and get a stock set. The same goes for grooved main-bearing journals—this hot-rod modification does no good. You don't want it at any price—*even free!* The only thing accomplished with these "tricks" is to reduce bearing area and weaken the crankshaft.

Bearing Size—The Federal-Mogul main-bearing set you should use is part number 4400M. These bearings are available in undersizes of 0.001, 0.002, 0.010, 0.020, 0.030 and 0.040-in. Chevy big-block mains have numerous different part numbers, depending on the year and undersize you need. Order your bearings from Chevy by year, undersize, and engine size.

To determine the correct-size bearings, you must compare your crankshaft's main-bearing-journal diameters to the stock diameter. Chevy lists a range of acceptable main-bearing diameters, 2.7485—2.7495-in. Use the average, 2.7490-in., for measuring purposes. Mike your crankshaft journals to see what they actually measure, then subtract that figure from the nominal size, or 2.7490 in. Let's say you miked your crank and found the smallest journal to be 2.7390 in. Subtract your reading from the nominal figure to find the amount undersize your bearings must be; 2.7490 − 2.7390 = 0.0100-in. You need to ask the parts man for 0.010-in. undersize—ten-under—bearings.

Use any of the first four journals for your measurements. The rear journal is slightly smaller than the other journals, so it will have somewhat greater oil clearance. Chevy lists a 2.7478-2.7488-in. range for the rear-main journal. You can measure this journal if you want—but adjust your figures.

Of course, that is the hard way to determine the correct bearing size. The easy way is to look on the card wired to your crankshaft by the machine shop. It will say what they did to the crank and you *should* be able to safely order bearings with this information; measure the journals just to be sure. If it reads .010 in large, scrawled, black felt-tip, that's the machinist's shorthand for 0.010-in., which is what he ground off the bearing journals, *both main and rod.* That's the undersized bearings you need.

There's a good chance the main journals were not ground, but the connecting-rod journals were. Then there will be two notations, such as, STD, (for mains), and .010 (for rods). The main journals are listed first, rod journals second, so order accordingly.

Rear Crank Seal—Thank goodness a *split lip* seal is used at the rear main. Look in your engine gasket set and pull the two seal halves from all the other gaskets. Don't confuse the rear main-bearing seal with the other rubber parts in the gasket set, including the manifold seals. Put the two seal halves on the bench where you can get at them and return the gasket set to your parts-storage area.

Wipe bearing bore and backside of bearing with a dry paper towel to remove any oil. Oil or dirt between bearing and its bore will inhibit bearing-to-block heat transfer and will distort bearing. Align bearing tab with notch at edge of bearing bore. Press bearing in place.

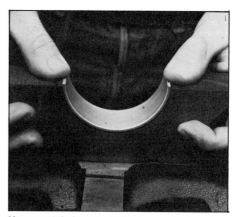

Ungrooved insert halves go in bearing caps. Use the same method to install them.

Bearing-Journal Size—Modern manufacturing techniques turn out marvels of precision mass-production. A good example are the bearings in your engine. There are all kinds of close tolerances that must be maintained when making bearings. When you multiply the tolerances by the thousands of bearings produced, you can get some idea of the problems involved. After noting all possible chances for error, it must be said that mistakes made in manufacturing bearings are rare.

As I said before, the biggest mistake is getting the *wrong bearings* in the *right box,* or vice versa. You can easily check this because the size of each bearing is marked on the backside, or shell, of the bearing. Sometimes standard-size bearings are not marked other than their part number, but they could have STD on their backsides. Undersized bearings should always be marked. Merely compare the bearing markings with the printing on the box to see if they agree. I recommend that you do this when you buy your bearings. Open the box on the parts counter and positively identify the bearings in the box before you leave the store! Inspecting your bearings *on delivery* can save so much time that not checking is downright foolish!

Bearing-to-Journal Clearance—There is something about installing new bearings without checking their clearances that should give you the jitters. Although problems with bearings are rare and you also double-checked the numbers versus the contents of each box, the machine shop could still have made a mistake when grinding your crank. This is especially true if you have been dealing with a

very large engine-rebuilding firm; your parts may have had work done on them that was meant for someone else's engine. This happens sometimes because work orders for two big Chevy cranks can get crossed somewhere and the *right work* is done on the *wrong crank.* Your crank may have only needed a 0.010-in. grind, but it may come back with a 0.030-in. grind. That could really affect bearing-journal oil clearance!

There are two ways of checking main-bearing clearance. The fastest and simplest is merely to install the bearings in the block and caps, oil the bearings, lay the crank in the block without the rear crank seal, then bolt the caps in place. If the crank spins easily by hand the clearances are probably close enough. Check for possible excessive clearances by trying to move the crank vertically; you shouldn't get *any* movement.

Read on for how to install the main bearings and crankshaft.

The second, and much more accurate method, involves the Plastigage strips. It also involves laying the crank in the block. Turn to page 116 for Plastigaging the crankshaft.

Plastigage Rod Bearings—If all you had to worry about was the main bearings, you could go ahead with laying the crank right now; but the rod-bearings must be checked just like the mains. There are two ways of checking the rods. By hand requires that the crankshaft be out of the block. Or, you could Plastigage the rod-bearings with the crank in or out of the block. If you forsee no problems with the rod clearances, you could lay the crank now and wait to Plastigage several rods until you install them in the engine.

Regardless of how you do it, don't move the rod while torquing the bolts—torque them 50 ft-lb; 67 ft-lb with 7/16-in. bolts. Although it's difficult, it is not impossible if you have a helper hold the rod while you torque it, then remove the cap. Use the same procedure as you did for Plastigaging the crank. Rod-bearing oil clearance should be 0.0007—0.0032 in. according to Federal-Mogul.

If you don't have any Plastigage or don't feel like using it, check the rods by hand.

Checking Rod Bearings by Hand—Lay the crankshaft on the workbench so one of the rod throws is up where you can work with it. Take the rod/piston assembly that fits that journal and oil the bearing. Install the rod with bearings and cap, and bolt it up. Rotate the rod around the crankshaft journal while feeling for any hang-ups, roughness or tight spots. The rod should turn effortlessly around the crank. Next, attempt to rock the rod on the crankshaft 90° to its normal plane of rotation. You should get no movement. The same is true of trying to twist the rod on its journal—there should be no perceptible movement.

Work with one rod at a time until you've checked all eight. You'll have to turn the crank to get at the other journals so don't do this work on a crowded bench. You could nick a journal.

Main-Bearing Installation—First you must have the block upside down so you'll be able to work unhindered on both the bearings and crank. Check the caps and their block mating surfaces for nicks or grit. Remove any, if found. Nicks that project up

Unless you're Plastigaging the crank, oil main-bearings and rear oil seal. Spread oil over bearing surfaces with your fingers. If you feel any dust or grit, wipe bearings clean and reoil them.

With a good grip on it, carefully lower crankshaft straight down onto the bearings. Be extra careful that your fingers don't get pinched at the rear of the crank. There isn't much room at that end, so a fingertip grip on the flywheel/flexplate flange is the best way to go.

can be removed with a file. Use a few strokes with a flat, fine-tooth file held flat against the surface. If a cap is being filed, set the cap on the file and slide it back and forth a few times. Don't forget to remove any filings.

Next, double-check the bearings for size by reading the numbers on their shells. All aspects of bearing installation require extra cleanliness, so use a paper towel soaked with lacquer thinner to wipe off the back of each bearing and its bore in both the block and cap. Don't leave any lint on the parts.

As for the bearing working surface, there are a lot of different ideas as to what's best to clean them. Frankly, a lot of them are pretty dumb, especially the old-wives tale about burnishing the bearings with steel wool, sandpaper or the like. You can bet the bearing manufacturers go to great lengths to provide the proper surface for engine break-in, and I don't see any sense in scratching away all their research.

The grooved-bearing halves go into the block. The non-grooved halves fit into the caps. Install each bearing by setting the plain end in the bore close to its normal position. Push down on the end with the small, bent locating tang to force the bearing into its bore. Using your thumbs, one on each end of the bearing, seems to work well.

The front four bearings install with little effort. The rear bearing takes somewhat more effort because its *thrust flanges* fit tightly over its bearing web and cap.

Install the bearings in both the block and in the caps. Remember when installing the bearings: Keep oil off their backsides. They must be absolutely dry and clean.

Skip Oil Seal—If you are installing the crank to check its journal-to-bearing oil clearances, bypass installing the rear-main oil seal. Leave the crank seal out or it will affect the readings by holding the crank off its bearings. To install the seal, turn to page 116.

Clean Crank—You must first make the crank ready for testing. As it comes from the machine shop it will be coated with grease or oil and may even have grinding grit all over it.

Wash the crank with *clean* solvent, removing all oil and grease. This will get all the accumulated dust and grime off the journals. If you find any dirt, clean the crank further until you are sure it is *all off.*

Pass a piece of welding rod or heavy wire through each oil passage. Be careful not to scratch the bearing journals while doing this. Then flow as much solvent as possible through the passages to remove loose particles. If you have compressed air, blow out the passages, taking care to shield your eyes. Finish up by drying the crank thoroughly with paper towels or compressed air.

Install Crankshaft—If you are preparing to Plastigage the main bearings, don't oil them. Otherwise, smear oil on the crank journals and grooved upper bearing halves in the block. Carefully lay the crank in the block. Keep the crank going straight down so the oil slinger at the rear won't hang up in its groove. Because the crank is so heavy, this job is best done by two people.

Now for the caps. Oil each lower bearing half as you go, spreading the oil with your fingertips—unless you're preparing to Plastigage. Observe the cap number and arrow so you will install the cap on the right bearing journal and in the proper direction.

If you are going to check oil clearance with Plastigage, read about Plastigaging before continuing.

Remember how the caps were tightly wedged in their registers during disassembly? This tight fit is important to main-bearing alignment and necessitates a little caution during bearing installation. Place each cap carefully in the block and tap it lightly into its registers. Use light pressure on the main-bearing bolts to seat the caps in their registers.

Torque all but the rear main-bearing-cap bolts. The proper way to torque main-cap bolts is a little at a time. Start with a ratchet and socket, running the bolts in until they are snug, or torqued about 10—15 ft-lb. Using your torque wrench, tighten the bolts in steps of 10 ft-lb. Begin at 20 ft-lb, then go to 30 ft-lb, then 40 ft-lb and so on. Final torque is a healthy 95 ft-lb for two-bolt caps; 110 ft-lb for four-bolt caps. Each bolt in a four-bolt cap gets the same torque.

Align Rear Main—An extra step is necessary when tightening the rear-main bearing because the thrust bearings need to be aligned. To align the thrust bearings, install the cap and run the bolts down finger-tight. Using two large screwdrivers, force the

Position main-bearing caps on *their* journals and in the right direction: cap numbers correspond to journals' and arrows point to the front. Light tap on the side seats cap in its register. After they are in place, double-check position and direction of caps.

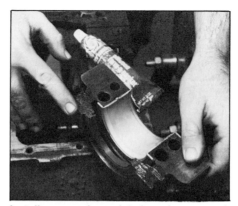

Install rear-main-seal halves in cap and block. Seal lip should point forward. Apply a light coat of silicone sealer to seal ends and cap as shown.

Torque main bearings in 10 ft-lb increments until you reach 90 ft-lb. Finish with 95 or 110 ft-lb for two- or four-bolt caps, respectively. Work with one cap at a time.

After installing all bearings and caps but the one you're checking, lay a strip of green Plastigage lengthwise on *dry* bearing journal. Install dry bearing and bearing cap and torque bolts 95 ft-lb, if it's a two-bolt cap, and 110 if it is a four-bolt cap. Do not turn crank.

Carefully remove bearing cap and measure flattened Plastigage with scale on sleeve. Plastigage should be the same width its full length. Consider 0.002—0.003 in. clearance to be very good.

crank back and forth several times, then hold the crank forward while you torque the rear main-cap bolts.

Spin Crank—Don't do the following if you didn't oil the bearings.

After all the caps and bearings are installed, spin the crank. Feel for any obstruction. The crank weighs about 70 lb, so don't expect it to spin like a top, but it should turn freely throughout 360°. If the crank binds at any point, something is definitely wrong and you should investigate thoroughly until you find what the problem is. First thing to check is your main-cap installation, then the bearings. Check each one.

Assuming the crank spins freely, it is now time to remove the rear-main cap and install the seal

Plastigage—A thorough Plastigaging job means checking all of the bearings. Journal must be wiped dry if

it was oiled, also bearings. Start with the front main and work towards the rear. Install all main caps except for the one you're checking.

Remove the strip of Plastigage from the sleeve and snip off a piece long enough to stretch the length of the bearing journal—front-to-rear. Don't worry about the fillet radius. And don't toss the sleeve away. You'll be needing it later. Install the main-bearing cap just as you did the other four, torquing its bolts 95 or 110 ft-lb for two- or four-bolt caps, respectively. *Don't turn the crank.* Doing so will smear the Plastigage, requiring you to remove the cap and start over.

Remove the cap and measure the flattened Plastigage with the scale at the edge of the sleeve. Federal-Mogul specifies 0.0005—0.0034-in. oil clearance for mains 1—4 and 0.0008—0.0038 in. for main 5. If clear-

ance is not within tolerance, recheck the journals and bearings. A too-wide strip means too little clearance; too narrow means excess clearance.

After measuring the Plastigage, wipe all traces of it from the journal and bearing. Reinstall the cap, then proceed to the next one to Plastigage its journal.

After you've checked all journals, remove the caps and crank. Install the rear-main seal; one half in the block and the other in the cap. Use the procedure described in the following section. Oil the bearings and seal, lay the crank back in the block and install the caps.

Rear Oil Seal—With all the crank checking done, the rear-main oil seal can be installed. Remove the crank from the block. Inspect the rear main-bearing area, cleaning out any oil that might have found its way into the seal groove. Install seal halves in both the cap and block so their ends are flush with the block and bearing-cap mating surfaces.

The lip of the seal must point toward the bearing—to the front of the engine. Reinstall the crankshaft and all caps except for the rear one. Before fitting the rear main-bearing cap, spread some RTV sealer over the ends of the seal and bearing cap to the rear of the bearing. See the photo. A relatively thin layer of sealer will do the job. Make sure the sealer is spread evenly and doesn't have "chunkies" of dry sealer in it or the cap may not seat properly. If you make a mistake, wipe off the old sealer and try again, it won't hurt anything. You can seal both the block and cap, or just the cap.

The object is to seal the ends of the seal without getting so much sealer

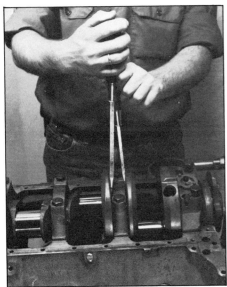

Thrust bearing must be aligned before torquing rear-main bolts. Force crank back-and-forth a few times. Then, hold the crank forward and snug bolts.

Checking crankshaft end-play with dial-indicator. With dial indicator positioned with its plunger square against crank nose, force crank to rear with a screwdriver; zero indicator. Force crank forward and read end play. End play can also be checked with feeler gages. Force crank in one direction or the other and measure end play with feeler gages between the thrust bearing and crank thrust face. Maximum end-play is 0.007 in.

between the block and bearing cap that the cap will not set correctly on the block. In addition, excess sealer may squeeze into the bearing area, causing possible bearing and bearing-journal damage. I put light coatings on both parts in my engines without any problems—just don't overdo it with the RTV.

Just as you did before, align the rear-main thrust flanges by prying the crank back-and-forth. Use two big screwdrivers as shown. Hold the crank forward and tighten the bolts to hold the bearings in place. Finish torquing the bolts.

Crankshaft End Play—You need to make one more crankshaft check; how much the crankshaft moves back-and-forth. This front-to-rear movement, or *walking,* is called *end play.* End play is controlled by the rear main-bearing thrust flanges, so if end play is excessive, the flanges are too thin. If end play is too little, the thrust flanges are too thick or the bearings are not aligned.

Machine shops and affluent enthusiasts check end play with a dial indicator set up parallel to the crank with its plunger against the crank nose. The crank is moved to the rear as far as possible with a large lead mallet or pried back with a large screwdriver. With the crank held in its rearmost position, the dial-indicator is zeroed. The crank is then pried forward as far as possible and end play is read directly from the indicator.

The second method is to move the crank just as you would with the dial-indicator, but measure end play with feeler gages. Move the crank to the front or rear and find the feeler gage that just slips between the bearing and crank thrust faces. Feeler-gage thickness is the end play. It is best to repeat the procedure several times to make sure you get the correct reading.

If end play is less than 0.005 in., you must remove the crank and both rear main-bearing halves. Remove enough material from the bearing front thrust flange to bring end play within specs by lapping it against 320-grit sandpaper set on a piece of glass. Butt the bearings halves together, just as they would be when installed, and lap them together.

It won't take much lapping, so be careful not to remove too much material. Make frequent checks by measuring the front flange with a 0—1-in. micrometer to make sure you don't go too far. I'm sure you'll be happy to hear that less-than-minimum end play is very rare. I really doubt that you will have any problems with it. Less than minimum end play could be the cause of a hard-to-turn crankshaft. However, if you've been having trouble in that area, tight end play could be one of your problems.

Maximum allowable end play is 0.007-in. If your crank has more end play than this, there isn't much you

can do about it except trade it in on a crank kit. There is *no oversized thrust bearing* available and adding material means building up the thrust-bearing surface and grinding it to the proper dimension. This is the type of work you should let a crank rebuilder worry about. If you really want to have *your* crank, talk to your machinist about welding it up.

End play is an important measurement. If you find your crank has an end-play problem, don't slap the engine together and hope the problem goes away. Chances are nothing really serious will happen, but excessive end play may allow the crank to move far enough back and forth that the rod and main bearings may edge-ride the bearings. Also, if you have floating rods, the pin bores are likely to wear rapidly. In extreme cases a rod could cock on the crank and break.

TIMING CHAIN

Although I didn't do it because I assembled my big-block on an engine stand, you can install the timing chain and sprockets now if you forsee a problem with the camshaft sliding out of its bearings. The chain and sprockets will hold it in place. Refer to page 123 to install the chain and sprockets.

PISTON-RING CHECKING & FITTING

You should have assembled the pistons and connecting rods by now. If you haven't, refer back to page 89 and do it now.

With the rods and pistons assembled, prepare and install the rings. Just like the bearings, today's rings are the result of millions of dollars of research and production progress. Very rarely will you have any trouble with piston rings, but you should still check them.

End Gap—The most important ring check is end gap. The important idea here is that the gap must not be so small that the ends of the ring touch or so wide that power is lost. If ring ends touch, the ring may break or score the bore. If the end gap is excessive, combustion-chamber pressure will be lost into the crankcase. The big Chevy seems to operate well with fairly wide gaps, so stay on the high side of end-gap specifications.

Checking End Gap—Measuring ring

end gap is easy. Remember the bore-taper measuring method that was computed from end gap, page 69? This is the same check, only simpler. Select a top ring—also known as the *first compression ring*—from the ring pack and place it in the number-1 bore. Normally you would use the top of a piston to push the ring down in the bore to square it up. However, if your pistons are the pop-up type, this won't work. Instead use a can that just fits in the bore. Or, a 6-in. machinist's rule with a slide clip used as a depth gage will work fine.

If you didn't rebore the block, push the ring down to the bottom of the bore. Bore diameter will be least there because of little or no bore wear and end gap will be smallest. Use your feeler gages to measure end gap.

A formula for figuring the correct compression-ring end gap states that gap should be 0.0045 in. for each inch of bore diameter. You can use this formula if you wish, or use the standard 0.010—0.020 in. for compression rings and 0.020 in. for oil rings. Oil-ring-gap specs vary, but if they don't exceed 0.030 in., they'll be OK. Measure the gap on all the rings for each cylinder. Besides verifying that the rings are correct, checking the end gap also double-checks the boring you might have had done.

Setting End Gap—If you find the ring gaps to be way too large, you probably have the wrong ring set. Exchange it for the correct one. If the gaps are too small, they can be opened up by filing. Now, before you begin looking for a big *bastard-cut*—coarse-tooth—file, let me explain that this is a delicate job. It is nothing you should shy away from, but you must have a file fine enough to do the job without cutting gouges in the ring end. What you need is a small *smooth, double-cut* flat file. Secure the file in a vise or clamp it to the edge of the work bench with a pair of Vise-Grips or a C-clamp.

By clamping the file, you'll have much better control. Pass the ring over the file, making all cuts from the outside of the ring to the inside. File one end only. Filing from the outside keeps a large burr from forming and Moly rings from chipping off at their outside edges. It will take very little filing to get the ring to size, so don't go too long without checking your progress. After filing a little, deburr the ring end and recheck its gap. It

Like bearings, you can get the wrong rings. Check ring end gap before installing rings on pistons. End gap should be 0.010—0.020 in. for compression rings; 0.020—0.030 in. is OK for oil-ring rails.

File one end of ring to increase gap. A special tool is available for this job, but a vise-held file works well. Hold ring square to file and file only in direction shown. Recheck end gap frequently because it will increase dramatically. Deburr ring end with a light stroke of the file on each of the four corners of the filed end. Keep rings in order so they will be installed in the cylinders they were checked in.

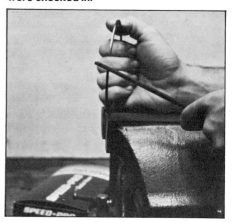

won't take long to get the feel of how much you're removing with each pass.

If you are lucky enough to be working in a machine shop, or just like buying tools, you should be aware of a special rotary file that grinds ring end gaps. It doesn't do a better job than you can with a fine file, it just does it a little faster and easier.

Finish the job by removing the sharp edges formed by your filing. The outside edge of the ring probably doesn't have a sharp edge, but the inner edge and sides will need a *light* cleaning up with a *very fine* file or 400-grit sandpaper. Go lightly, just remove the sharp edges so your finger cannot feel any roughness when you pass it over the ring end. This will take just a touch or two with your fine file or sandpaper.

Check and gap one ring set at a time, then move on to the next one. Keep each set of rings with the bore that it was measured in, or the gaps will come out incorrectly. The paper pack the rings came in is an excellent container for keeping the rings organized; simply mark the cylinder number on the package.

Also, make sure you do not confuse the top and second rings in the set. They look alike, but are different. By working with only one ring at a time you can avoid mixing them up. If you do get mixed-up, check the marking near the end of the ring. These marks are there to designate the top of the ring and are usually different for the top and second ring and, in case of a truck, the third ring.

INSTALL PISTON RINGS

Now that the rings are gapped, clean them thoroughly before you put them on the pistons. I've already mentioned the ring expander. If you don't have one by now, I suggest you get one. You'll be doing yourself and the engine a big favor.

Steady the rod/piston assembly in a vise. This is about the only workable way of holding the piston motionless so you can install the rings. When you clamp the rod in the vise make sure you are using some sort of *soft jaws*, such as two pieces of wood on each side of the rod. Never clamp a rod directly in a vise. You'll cut a sharp groove in the beam, creating a starting point for a stress-induced crack. Using a rag between the vise jaws and rod does little good, so use wood blocks or store-bought soft jaws. Position the assembly so the piston skirts sit on top of the jaws. This will keep

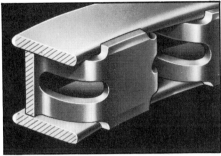
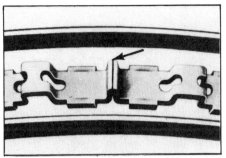

Three-piece oil rings have two scraper rails installed over an expander/spacer. It is essential that the expander/spacer ends butt (arrow), not overlap. Otherwise, you'll end up with a scored cylinder wall. Drawings courtesy of Sealed Power Corporation.

Speed-Pro oil-ring expander/spacer has a small plastic block bonded to each end. Blocks help keep ends from overlapping.

After placing expander ring in bottom groove, wrap rail into oil-ring groove, above expander/spacer. Keep your finger over the free end of scraper so it doesn't gouge piston. Install bottom rail and adjust end gaps according to nearby drawing.

Expanding compression rings with the thumbs can be painful, or nearly impossible. Although you can install compression rings this way, don't be surprised if you break one.

If you don't have a ring expander, get one. An expander is well worth the money, particularly if it keeps you from breaking a ring. And make sure the dot or *pip mark* is up. Position end gaps.

the piston from wobbling back and forth. Don't clamp the piston!

If you don't have a vise, buying one would be money well spent, even long after your big-block is back on the street. If you can't see buying one, secure the rod to a bench with a C-clamp or have a friend hold the rod/piston assembly while you install the rings.

Install Oil Rings—Rings are always installed from the bottom up, so the oil ring goes on first. As you know, the oil ring is actually a three-piece assembly: two side rails and an expander/spacer. Each piece must be installed separately. Study the ring-placement chart so you'll get the end gaps in the right place before starting.

Select the oil-ring expander/spacer, the flimsy, wavy ring. Fit it in the bottom ring groove with its ends located in the small tab within arc A on the chart, page 120.

If you are using Chevy rings, you'll find a tang on the back of the expander/spacer. This projection fits into a hole in the ring groove within arc A. Some aftermarket rings don't have this tang, just position their end

gaps within arc A. All oil-ring components are light and handle easily, so you won't need the ring expander for these.

Next on is the top rail, or scraper. This thin ring fits between the expander/spacer and the groove. Because the rail is so flexible, you can fit one end in the groove located on arc B and hold it there with your thumb. Then the rest of the scraper can be threaded into place by running your finger around its edge.

The object here is to keep from scratching the piston with the end of the scraper. Avoid scratching the piston by holding the end of the ring away from the piston with your thumbnail. Sometimes you'll get your thumbnail caught between the piston and ring, but that's better than scratching the piston.

Install the bottom scraper in the same manner.

Once the bottom scraper is in place, check that the ends of the expander/spacer are not overlapping. See the accompanying drawing. Some expander/spacers have plastic blocks on each end that prevent this.

Remove the scraper and start over if you find that the expander/spacer ends are overlapping. If you don't, severe bore damage will result.

Install Compression Rings—Although they appear to be simpler than the oil ring, compression rings have hidden design features that require attention and care on your part. For example, an oil ring has no top or bottom; compression rings do. The top and bottom are a result of the *twist* built into a compression ring. Although twist is virtually invisible, it is there to help the ring seal to the ring grooves.

So you'll know which side is up, the top side of a compression ring is marked with some sort of dot or indentation near the gap. This mark must face up when the ring is installed on the piston, otherwise the twist will be going the wrong way. That could result in excessive blow-by, power loss, increased oil contamination and a host of other bad things. Besides my warnings, you'll find an instruction sheet with your rings. Read it carefully and follow it to the letter. You'll have a better running engine if you do.

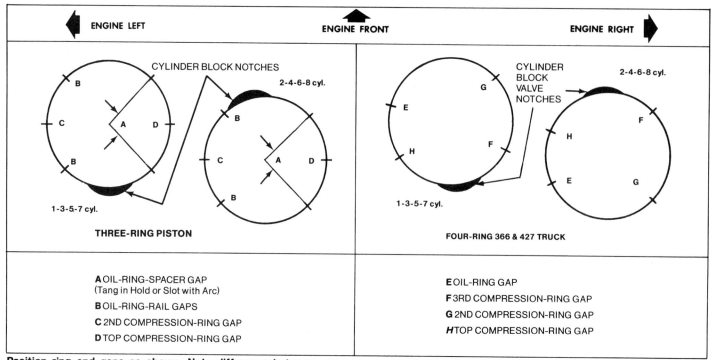

Position ring end gaps as shown. Note difference between passenger-car and light-truck three-ring pistons and four-ring 366- and 427-truck pistons. Drawing courtesy of Chevrolet.

Spreading, or expanding, the compression rings is the difficult part of installing them. They are very rigid and, worse yet, brittle. If you spread or twist one a little too much you'll have a two-piece ring on your hands. Ask any mechanic—no doubt he can tell you about a ring that broke during installation for seemingly no reason at all. All these horror stories are true of installing rings by hand.

Using a ring expander reduces risk of ring and piston damage to practically zero. This is especially true when you use the type that cradles the entire ring. Even the inexpensive style, which only grasps the ring by the ends, is way ahead of doing it by hand. The key idea in installing rings is to not expand them any more than necessary. Spread the ring just enough to clear the piston, move it to its groove, then relax it into position.

Go gently when installing the compression rings. Use both hands to guide the ring over the piston. Don't cock the ring on the piston because this makes it necessary to expand the ring further, increasing the chance of breaking it.

Install the bottom compression ring first. When the second compression ring is installed—third in the case of a truck—install the next one up. If you get the top ring on first, remove it and install the bottom one. There is no

way you'll get one compression ring over another without breaking it.

Work with one piston at a time, installing all the rings. When you are done, check the marks to make sure the compression rings are installed top-side-up. Then position the gaps according to the drawing. Set the finished piston aside and start with another. The rings will move around somewhat as you handle the rod/piston assemblies, so recheck the ring-gap spacing immediately before installing each piston in its cylinder.

If you are installing compression rings by hand, there are some tricks you should know about. First, wrap your thumbs with tape. This will protect them somewhat and allow you more control over the ring. Take the second compression ring—third with a truck— and hold the ends with the end of your thumbs. Spread the ring and guide it over the piston with the ends slightly lower than the rest of the ring. Gently spread the ring *with your thumbs only.* Do not allow your fingers to press against the outer edge of the ring; they will act as a fulcrum and the ring will break before you know what's happening. Use your fingers to guide the ring downward instead. Pass the ring over the piston to its groove, then relax your thumbs. Install the next compression ring after your thumbs revive.

PISTON & CONNECTING-ROD INSTALLATION

To install the pistons and rods without damage, you'll need several special tools: a ring compressor, crank journal protectors that slip over the rod bolts, an oil can, a large can half-full of motor oil, and a hammer. You'll use the hammer handle to push the pistons down their bores. The large can of oil is for dipping the pistons into before installation. If you can't find a suitable can, you can do the job without it, but it is the easiest way to oil the piston thoroughly before installation.

Get Organized—Piston installation goes a lot more smoothly if everything is laid out before hand. All your tools must be clean and within reach. The block must be positioned so you can work the piston in from the top and guide the rod with your other hand from the crankcase.

If you have an engine stand, turn the block so one cylinder bank points straight up. This will let the rod hang without touching the bore and makes it easier to guide. If you are working on a bench, turn the block over so one of the banks projects over the edge. You'll find that it is actually easier to have the block on its back so you'll have access to the connecting rod. It is less difficult to push the piston up its bore than to connect the

Tools needed for installing pistons and rods: hammer, ring compressor, squirt can, bearing-journal protectors, and oil and a coffee-type can to put it in. In addition, have some clean rags and paper towels handy.

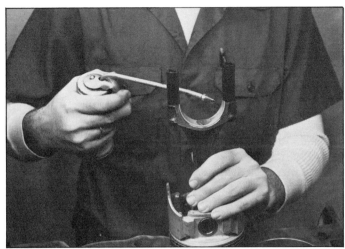

Double-check ring end-gap positions, then oil rod bearing. Using a squirt can is a lot less messy than dipping big-end of the rod in oil. While the oil can is in hand, oil cylinder wall.

rod and crank upside down.

Clean the Bores—Wipe the protective oil coating out of the bores with paper towels. The towel should come out clean. Use your squirt can to put a fresh coat of oil on the cylinder walls. Spread the oil around with your fingers. If you feel any grit, now is the time to clean it out. Inspect the pistons, rods and your tools for dirt and clean them as necessary with fresh oil. The rod bearings should be brought out of their boxes and partially unwrapped.

The rod/piston assemblies should be arranged in order next to the block so you won't have to walk around searching for the right piston. To do one cylinder bank at a time, have all the odd-numbered pistons in one row and all the even-numbered pistons in another.

As soon as everything is arranged you will be ready to install the pistons. Because of the cleanliness required and the numerous steps that must be followed in proper order, I suggest that you arrange them so you can install all the pistons in one session. That way you won't skip any steps and the parts will stay cleaner.

Position Crankshaft—Go through your collection of parts until you find the crankshaft-damper bolt and washers. After cleaning the bolt and washers, install them in the crank nose.

You'll now be able to use a wrench on the damper bolt to turn the crankshaft. This is necessary so you can turn each connecting-rod crank journal to BDC. That way you'll have room to reach the rod as you guide it

onto its journal. As you start out you may not need the bolt, but as more pistons are installed, the drag will increase until you must use a wrench to turn the crank.

Ready Rod/Piston—To install a rod/piston, first turn its crank journal to BDC. Remove the cap from the connecting rod. Wipe the rod and cap bores clean and dry with a paper towel, then wipe the back of the bearing. Press the bearing inserts into place in the rod and cap with your fingers and smear clean oil over them. Slip the journal protectors over the rod bolts—this is a must. If you don't have the ready-made protectors, use two lengths of 3/8-in. hose.

Next, coat the rings and pistons with oil. You can do this with a squirt can. However, a large container half-filled with oil is a lot easier. Just turn the rod/piston assembly upside-down and dunk the piston in over the rings and piston pin. Move the rod back and forth to work oil in between the pin and pin bore. Lift the piston out and let the excess run off into the can. Don't worry about the mess. You can clean it up after you have all the rods and pistons installed.

Squirt some more oil onto the rod bearing. Spread the oil evenly over the bearing surface.

Compress Rings—I've already mentioned the two common types of ring compressors—plier and cylinder type. The plier type is preferred because it is a bit easier to use. Some mechanics claim it does a better job of compressing the rings. You compress the rings by squeezing the plier handles while a ratchet device keeps the compressor

Install rod bearings in a manner similar to how you installed the mains. Wipe bearing halves and bearing bore clean with a paper towel. Press bearing halves into cap and rod.

Next on are the journal protectors, one on each rod bolt. Not only will they protect the crank and cylinder wall, they'll also keep the bearing in place.

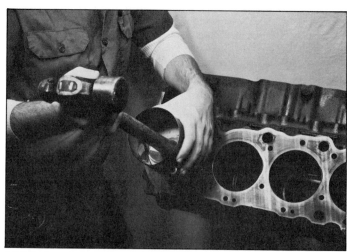

With piston and rod positioned correctly—piston dot and rod number to the outside—insert rod into bore. Take care not to bang rod into cylinder wall. Tap ring compressor flush onto deck using hammer handle.

Lightly tap piston into its bore. If you feel any resistance, *STOP*. Remove piston and start over. What happened is a ring popped out from under the compressor before entering the bore. If you force the piston, you'll damage the ring and ring land.

Immerse piston in oil over the rings and piston pin. Let the excess oil run off. Wrap ring compressor around piston and compress rings.

from back-tracking. You release the compressor by tripping a latch on the ratchet.

Cylinder-type compressors are usually tightened with a large, square key. They too have the ratchet device that is released by tripping a small lever. Both types of compressors have a limited range of travel, particularly the plier type.

The plier-type ring compressor is limited by the metal band that fits around the piston. If the band size is incorrect you will need another size. Sealed Power's MT-117G 4-1/8—4-3/8-in. band will work for most big-blocks. For the smaller-bore 366, you'll need the MTF 3-7/8—4-1/8-in. band. Either will work with the 396 or 427.

You may have noticed that the ring compressor has a top and bottom; there should be an arrow on the cylinder or band indicating the bottom. The ratchet or handle attachment is closer to the bottom.

Fit the relaxed, cleaned and oiled compressor over the piston so the bottom is toward the connecting rod. Begin to tighten the compressor after centering it on the piston; the piston should just peek out the top of the compressor. As the compressor tightens around the piston, wiggle it back and forth on the piston so the rings will not bind in the compressor. You'll feel the compressor relax as the rings compress. Tighten the compressor until it is firm against the piston.

Double-check the number of the rod/piston assembly before placing the rod in the cylinder. Also make sure it is correctly aligned in the block. Numbers on the rod face go to the outside. The dot on cast pistons faces the outside of the block also. Forged pistons have no mark on their domes, but the pop-up portion gives you an excellent clue as to its correct orientation.

With closed-chamber heads, the greater of the two flat areas on the piston dome goes toward the center of the block, closest to the intake manifold. Open-chamber pistons are different; they have only one flat area on the dome in most cases. It also goes toward the center, toward the intake manifold. Flat-top pistons must be aligned using the intake-valve relief as reference. Some aftermarket pistons have no marks and can be used on either side of the engine and, if they are flat tops, either way on the rod *if there is no pin offset.*

Install Rod/Piston—Guide the rod into the cylinder with your free hand. Don't let the rod bang against the bore on its way down. The piston skirts will slip easily into the bore until the ring compressor butts against the deck.

Force the compressor down so it butts squarely against the block deck. Do this by tapping around the upper end of the compressor with your hammer handle after the compressor has butted against the deck. Go lightly, and watch the band of a cylinder-type compressor line up where it overlaps as you tap it down.

Use your free hand in the crankcase to guide the rod. Sometimes the piston will turn on its way in and the big end of the rod will hang up on a crank counterweight or the block. If so, turn the rod/piston assembly so it lines up with its journal.

Check the ring compressor to make sure it is still tight around the piston. Then use the hammer handle to *tap* the piston into its bore. If the piston "goes solid" and doesn't want to go any farther, **STOP!** A piston ring has popped out of the ring compressor and has snagged just before entering its bore.

If you hammer hard on the piston in a effort to force the piston, you'll break a ring or ring land. If you break the ring and continue your installation, the broken ends of the ring will scratch the cylinder. You won't notice it until too late, and then you'll really be mad.

There's no big deal to stopping and pulling a rod/piston assembly from its bore. Merely inspect the rings, refit

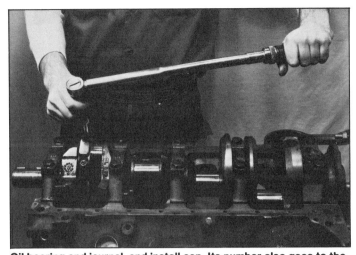

Oil bearing and journal, and install cap. Its number also goes to the outside; oil squirt hole goes to the inside. Install nuts and snug them for now. Torque them after all pistons and rods are installed. Torque 3/8-in. rod nuts 50 ft-lb; 67 ft-lb for 7/16-in. rod nuts.

Brass drift is ideal for driving crank sprocket on. Stop about 1/2-in. short of driving sprocket "home," then install chain and cam sprocket. Once cam-sprocket dowel and bolts are engaged and timed correctly, finish driving on crank sprocket.

the ring compressor and start over.

When the last ring enters the bore, the ring compressor will relax and slip off the piston. Set it aside. Push the piston gently down its bore with one hand while guiding the rod big end with your other hand. Double-check that the bearing and journal are oiled, then fit the rod to the journal. Remove the two journal protectors.

Install Connecting-Rod Cap—Place the rod cap on the rod bolts only after you have made doubly sure it is the correct cap and is aligned properly. It's easy to keep the rod caps straight if you work with one rod at a time, so have only one cap unbolted at a time. The bearing must be well oiled, of course. Don't use moly grease on the bearings, it isn't needed. Remember, the rod and rod cap are both numbered. The numbers should be on the same side of the rod, pointing away from the camshaft. The bearing tangs should also be on the same side of the rod.

Run the rod nuts on until finger tight. Then torque them to 50 ft-lb if you have 3/8-in. bolts, 67 ft-lb if you have 7/16-in. bolts. Torque each bolt 10 ft-lb at a time, alternating between bolts. Torquing one bolt all the way, then the other, will distort the bearing and cause trouble.

So, one rod/piston assembly is in the block; there are only seven more to go. Use a wrench on the damper bolt to turn the crankshaft to bring the next journal to BDC. Install the remaining pistons the same way, and remember to use the journal protectors over the rod bolts each time. If you don't, you are risking a big nick in

your crankshaft. When you have finished with the first cylinder bank, you'll have to move the block to expose the other cylinder bank.

CAMSHAFT DRIVE

Your cam may either be driven by chain and sprockets or gears. I'll first explain how to install the more common chain-and-sprocket setup.

Chain & Sprockets—Install the crankshaft sprocket first. If the Woodruff keys—those half-moon pieces that fit into the crank snout—are missing, you'll have to replace them. Ask at the machine shop where the crank was machined; they'll probably give you the two you need. Tap the keys into place, then slip the crank sprocket over the crank snout and into engagement with the keys. You'll need to tap the sprocket in place with a drift. Don't hammer in one spot, but work around the sprocket. The sprocket will "walk" onto the crank snout until it bottoms against the front main-bearing journal.

If you are using a stock crank sprocket it'll only have one keyway. Cloyes True Roller crank sprockets have three keyways, one that provides stock cam timing, another that retards the cam 4° to the crank, and another for advancing it 4°. These different keyways allow easy cam-timing adjustments for high-performance engine builders. You should install your stock cam *straight up*—neither retarded nor advanced—if you're building your engine to stock specs. The keyway for standard cam timing on the Cloyes sprocket is marked with an O. Place that keyway over the wood-

Correct relationship between crank and cam sprockets. Dowel hole in cam sprocket is at three o'clock and its timing mark straight down. Crankshaft-sprocket mark is straight up and crank key is at 2 o'clock. Cloyes crank sprocket has three keyways for different cam timing.

ruff key during installation.

Connecting the camshaft and crankshaft together is a matter of correctly positioning both the crank and the cam. To get the two *in sync* you must start by turning the crankshaft to cylinder-1 firing position. This is TDC on cylinder 1. With cylinder 1 on TDC, note that the crankshaft keys are not straight up, but a little to the right as viewed from the front of the engine. What will be straight up is the small O on the crank sprocket. You can position the crankshaft more accurately than you think if you watch the

Torque cam-sprocket bolts 30 ft-lb.

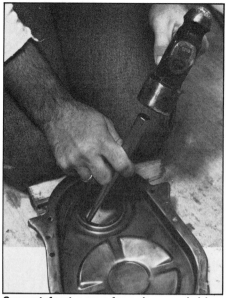

Support front cover face down and drive out seal. Take care not to distort cover.

Turn front cover over and back it up underneath seal bore. Apply sealer to seal OD. Position seal over seal bore with its lip pointing down, or toward cover. Back up seal with driver that contacts full face of the seal. Drive seal straight in until it bottoms; don't let seal cock or it may distort and leak.

piston as it approaches TDC. Run the crankshaft through several times to make sure you have cylinder-1 piston setting at TDC. The keyway will be at the 2 o-clock position and the 0 on the sprocket will be straight up, exactly between the crank and the cam.

To position the camshaft, turn it so the sprocket-drive pin is at 3 o'clock. Fit the camshaft sprocket over the cam and run one bolt in finger-tight. Examine the sprocket edge and you'll find a triangular-shaped mark near the 6-o'clock position. This mark must be between the cam and crank, directly in line with the crank-sprocket 0 for it to be correctly aligned. After you have turned the cam sprocket to this position, both marks on the sprockets will be opposite each other.

Install Chain & Sprocket—Because of the shallow-tooth design of the big Chevy camshaft sprocket, you can install the chain and cam sprocket with the crankshaft sprocket already in place. Use your left hand to drape the new chain around the crankshaft sprocket. While holding the chain up, slip the camshaft sprocket inside the chain. Before meshing chain and sprocket teeth, turn the cam sprocket so its triangular mark is just opposite the crankshaft-sprocket 0. You'll be able to mesh the sprockets together with the chain when the marks are correctly aligned.

Now lift the camshaft sprocket up until it meshes with the chain. Move the chain, not the sprocket, if you have to make any adjustments. Check the timing marks again to make sure

you didn't lose their exact relationship, then place the sprocket against the nose of the camshaft. The sprocket and camshaft should align closely enough so that you can install at least one sprocket bolt.

If the cam is somewhat out of position, move it with your fingers, then place the camshaft sprocket over the drive pin. Run the three sprocket bolts into the camshaft finger-tight. Stand back and eyeball the timing marks one last time. They should still be lined up. If not, remove the sprocket and start over.

When you are satisfied that the timing is accurate, torque the three camshaft bolts 20 ft-lb. If you want a bulletproof installation, remove one bolt at a time, make sure it's clean and dry, then give it a dab or two of Loctite instead of the usual oil. Now you can torque them.

Cam Gear—There's not much to installing the crank gear of a gear-driven cam; the cam gear is already in place. Line up the crank gear with the key in the crank and drive it on the nose of the crankshaft with a brass punch and a hammer until it is close to the cam gear. With its timing mark, or dimple, directly in line with the cam, rotate the cam so its gear timing mark will line up with the one on the crank gear when they are fully engaged. Finish installing the crank gear by driving it all the way onto the crank.

FRONT COVER & CRANK DAMPER

Clean the front cover—*timing cover,*

if you wish—if you haven't already done so. Scrape off all the old gasket material and check the gasket surfaces for flatness. If the gasket surfaces are not flat, use the hammer-and-punch flattening trick on them. You'll also want to change the oil seal. The seal is removed by driving it out from the inner—timing-chain—side. Use a punch to reach around the timing cover to the inner lip of the seal.

Suspend the cover on wooden blocks so the seal will be free to pop out. Work around the perimeter of the oil-seal backside with the punch to drive it out; don't hammer in just one spot. Use light blows so you won't distort the front cover. To install the new seal, coat its outer periphery lightly with a little non-hardening sealer. This will prevent any leaks between the seal and the cover.

Lay the front cover on the bench with its flanges down, then lay the seal over its bore. The metal shell faces up, the seal lip points down, into the timing cover. Now you need a tool, such as a very large socket, that approximates the seal diameter. This tool must be smaller than the timing-cover bore and larger than the oil-seal ID.

If you are very careful, you can install the oil seal without a tool. A medium-size punch and hammer are all you need, but you must be very careful to work around the seal and not cock it. The seal must go in straight, which is the advantage of using something like a large socket. By whatever means, tap the seal in

Run sealer around front-cover flange. Circle bolt holes with sealer. Fit gasket to cover and give it the sealer treatment, too.

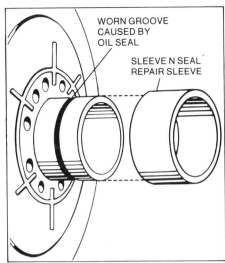

Install front cover to block, fitting cover and gasket over locating dowels. Thread in bolts and snug them down gradually. Final torque bolts 7 ft-lb. Bolts will require re-torquing as gasket relaxes.

WORN GROOVE CAUSED BY OIL SEAL

SLEEVE N SEAL REPAIR SLEEVE

If damper seal surface is severely grooved, repair sleeve will provide a new seal surface. Low cost of sleeve will also save high cost of new damper. Drawing courtesy of Fel-Pro Inc.

place. When the seal bottoms, you'll sense it. The whole timing cover will make a solid "bonk" noise when you tap the seal.

Turn the cover over and check the rear edge of the seal shell; it should be bottomed in its bore. If you cocked the seal, remove it and install another. A cocked seal will leak oil every time, and hammering a seal out and reinstalling it will further distort it. Work slowly and you will not cock the seal. If you do, buy another.

Front-Cover Installation—You'll find the timing-cover-to-block gasket in the gasket kit. Seal it to the timing cover with gasket sealer. Then apply sealer to the block side of the gasket. Timing-cover bolts should be clean and their threads lightly oiled. The block surface that mates with the timing-cover gasket must be clean and oil-free or the gasket will not seal. Use solvent or lacquer thinner to clean it. Place the cover on the block and install the bolts.

Prepare Damper—If the damper has not been cleaned or checked, do it now. After the solvent bath, test the integrity of the rubber. Place the damper nose down on the floor, stand on its center with the ball of your foot, grasp the outer ring, then try to move the outer ring relative to its center. Any noticeable movement means the rubber bond has failed and the damper must be replaced.

If the outer ring is solid, inspect the inner and outer mating bores for imperfections. Smooth any nicks, burrs or scratches with 400-grit emery cloth. This is the same abrasive cloth you used on the crankshaft, and it will give the damper oil-sealing surface some *tooth*. This ever-so-slightly

rough surface will retain just enough oil to lube the oil seal.

If the damper is OK, except for a deep groove in the oil-seal surface, you can recondition it with a *sleeve*. This sleeve fits over the damper nose and provides a new surface for the seal to run against. Buy this sleeve at your local parts house. It will be about 0.030-in. thick, not enough to distort the oil seal.

Make sure there are no burrs on the damper seal surface, then coat it with silicone sealer (RTV). Place the damper upside down on something that supports its center, then set the sleeve on the damper shaft. Place a block of wood squarely against the sleeve. Tap the sleeve in place, making sure it is starts straight. After the sleeve seats, wipe off excess sealer.

Damper Installation—Start the damper installation by oiling the damper-seal surface and the seal. Position the damper on the crank nose and rotate it until its keyway engages the key. It will be difficult to get the damper started by hand because of the tight fit between damper and crank. The factory-correct way to install a big-block Chevy damper is with a special tool that pulls the damper onto the crank snout with a large screw. A gear puller that has a center bolt that threads into the nose of the crank will work backwards to accomplish the job.

If the special tool is not available, go to Plan B: Drive the damper on with a large lead or brass mallet, or use a large socket against the *inner ring*. Hammering the damper on is not recommended because it can damage the damper rubber ring. This could

If you don't have anti-seize compound, oil crank snout before installing damper. Also oil damper seal surface. Fit damper to crank nose, rotating it until keyway engages key. Push damper on as far as it will go, then draw it on with a long bolt, nut and washer or drive it on. Do not drive against damper outer ring.

cause it to fail later. If you do hammer the damper on—like most engine rebuilders do—do so with many light blows, not a few of "Thor's Best."

After the damper is on far enough so its bolt will engage the crank several turns, use the bolt in combination with the hammer and socket to move the damper on the rest of the way. After the damper seats, torque the bolt 85 ft-lb. Keep the crank from rotating by wedging a hammer handle or 2x2 between the block and a crank counterweight.

OIL PUMP & PAN

Now that the timing cover is on the block, you can install the oil pump and pan.

Oil-Pump Drive Shaft—The oil-pump drive shaft in the big-block

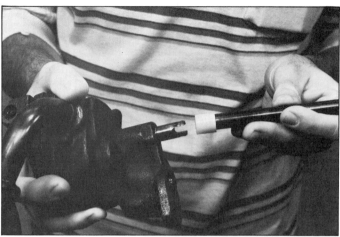

New oil-pump-to-drive shaft sleeve is a must. If you haven't done so, replace the old one. Sleeve goes on with a light push. Make sure connector snaps onto both oil pump and drive shaft.

Thin bead of *non-hardening* sealer ensures that oil pump won't suck air rather than oil. Do not use RTV sealer.

Install damper bolt, flat washer and lock washer. Torque bolt 85 ft-lb. Keep crank from turning with a short 2x4 between a crank counterweight and block.

Chevy is plenty strong, but the old plastic connector between the drive shaft and oil pump should be replaced; it will probably break if reused. If this happens, the oil-pump drive shaft will drop into the oil pan if the distributor is ever removed for service. Ask your Chevy parts man for part 3764554.

If you have a Corvette, you'll probably never have to pull the distributor out. But the distributor is so badly buried in almost every other GM chassis that it must be removed before a tune-up. If the drive shaft were to fall out of place, you'd have to remove the pan to replace it. In most chassis the engine would have to be removed first. Even worse, the engine will run

Fit oil pump to rear main cap and dowel. Install single bolt and torque 65 ft-lb.

without the drive shaft—and without a functioning oil pump. If the driver missed the oil-pressure warning light or didn't watch the gage, the engine could be ruined.

Install Oil Pump—One of the important things to accomplish before starting your engine is to prime the oil pump. Oil-pump priming is *not automatic,* even though the pump turns and the pan is full of oil—the pump may not suck *any oil* up and push it through the engine! This could happen if the oil pump were assembled completely dry. The gears and chamber walls would not seal and pumping efficiency could be so low that the pump couldn't prime itself.

To avoid this problem, make sure there is oil in the pump *before installing it.* If you still have the oil can used for dunking the pistons, use it again. Or you may need a wider container. Submerge the oil-pump pickup in oil and turn the drive shaft clockwise. You may notice on the first few turns that there is no load—the pump seems to freewheel. That's before priming starts. You can imagine the damage

that would occur if your new engine were being cranked and the oil pump was just spinning in thin air. It won't take much to prime it though, and after a few turns oil should gush from its top—the pump is now primed.

Let excess oil drip off and out of the pump and install it. Turn the pump upside down and let the excess oil spill out of the chamber, or it will surprise you when you invert it to install it. Wipe excess oil off the the outside of the pump, especially from the mounting surface.

Use the *new* plastic connector to join the pump and drive shaft. Give the two a tug to make sure the connector is fully engaged. Before bolting the pump in place, run a thin bead of non-hardening sealer around the pump mating flange on the rear main-bearing. This will seal the pump base; a good idea because there is no gasket and a good seal here is important from both priming and pressure standpoints.

Don't use RTV to seal the oil pump. You can wipe the excess off the outside of the pump, any that squeezes out on the inside can't be wiped off. It could block oil flow or clog passages.

After fitting the pump to the rear-main cap, install the one 7/16-14 bolt and torque it 65 ft-lb. Although it may sound stupid, make sure the oil-pump pickup is securely attached to the pump and pointing in the right direction. This may sound very obvious, but if the pickup isn't in place—you'll be ready for another rebuild after a few short minutes of engine operation!

Install Oil Pan—This is your final

Seal rear-main pan gasket with RTV sealer. Ends of this gasket overlap composition pan gaskets at each end.

Seal front pan gasket to bottom of front cover. As shown, it also overlaps ends of composition pan gaskets. Make sure rubber pegs engage holes in front cover.

chance to inspect the bottom end, so take a slow, careful look around. Check all the main- and rod-bolt torques. I also like to check oil-pump-bolt torque before buttoning up the bottom end—you should do the same. Also check the rods to make sure they are not cocked on the crankshaft. Merely grasp the big end of each rod in your fingertips and move it back and forth, front-to-rear, a little. You may find a rod that "pops" loose; it could have been slightly cocked on its journal. This cocking won't "throw" a rod, but it can harm the journal. After saying goodbye to the bottom end for another 100,000 miles, it is time to bolt on the pan.

Go to your gasket set and dig out the new pan gaskets, and the front and rear rubber pan seals. You'll also need the pan bolts. Start with the two end seals.

First, install the rubber seal at the rear, draping it over the rear main-bearing. It fits into a groove in the rear main-bearing cap and runs down to engage the gaskets at the pan rails. Coat this seal lightly with RTV. At the front of the engine, bond the rubber seal to the bottom of the front cover with RTV. Make sure the projections on the seal fit into the timing cover. Now lay the two pan gaskets on the block pan rails, sealing the gaskets to the block, not to the pan. Now, if you have to remove the pan, the gasket may not tear and you can reuse it.

The front and rear seals overlap the pan-rail gaskets where they meet. Don't think you have a mismatch, it's normal. Use RTV to seal the gaskets to the end seals. Squeeze some RTV

Seal composition gaskets to block pan rails with RTV, particularly at the four corners where gaskets overlap.

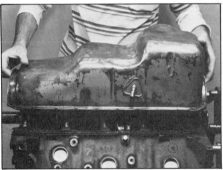

Some front-to-rear movement is necessary to get the oil-pan baffling over oil-pump pickup. Set pan squarely onto gaskets. Install and tighten pan bolts 14 ft-lb; 7 ft-lb for those that thread into front cover. If you have a more-accurate in-lb torque wrench, torque the bolts 168 in-lb and 84 in-lb, respectively. Continue torquing bolts until gasket is fully compressed.

between the gaskets in the small well provided in the rubber end seals.

Keep an eye on the gaskets and seals as you set the pan onto the block because they shift out of place easily. Position the pan to the block by starting a little to the front of where you want the pan to end up, then by lowering it and moving it to the rear. The front-to-rear motion should allow the internal baffling in the pan to clear the oil-pump pickup.

Go smoothly and don't drag the

pan across a gasket or seal. This will dislodge it. Once the pan is in place, start and thread in the 23 pan bolts. Now is an excellent time to use your speed handle. Run the bolts down until they are snug. Install and snug them from the center of each side of the pan, and work outwards. Alternate between sides.

Cork oil-pan gaskets relax after they are compressed by the tightened pan bolts. The problem is the bolts aren't tight anymore. Continue snug-

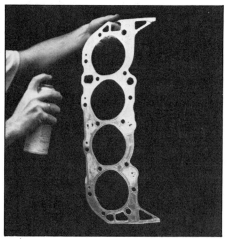

Unless instructions accompanying gasket set says otherwise, give each a silver paint job with high-temperature silver spray paint. An excellent alternative is to use High Tack or Copper Coat head-gasket sealer.

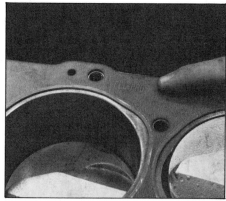

Look for UP and FRONT designations and position head gaskets on block decks accordingly. If those words don't appear on the gaskets, trade name goes up. Match gasket water-passage holes to those in block. Fit gaskets over the two locating dowels in each block deck. If dowels were removed, replace them before you proceed.

Now that you've recovered from installing crank, it's time for the heads. Lift each head and lower it squarely on to deck. Don't set head on gasket until dowels are engaged, then set it down squarely.

Short bolts thread into bottom row of holes; long ones go adjacent to exhaust ports and in valve and rocker area. If they aren't clean, clean threads with a wire brush. Apply anti-seize compound or non-hardening sealer to the bolts to keep coolant from seeping past threads.

ging them until they stay snug, then torque the 5/16-18 pan bolts 14 ft-lb; torque the two or three 1/4-20 pan-to-timing-cover bolts 7 ft-lb. Over-torquing these bolts will distort the timing cover and start an oil leak, so don't get carried away.

CYLINDER HEADS

Things are really starting to move along. The bottom end is done, the front of the engine is taking shape and now it's time for the heads.

Gather up those things you'll need to install the heads: bolts, gaskets, a can of high-temperature aluminum spray paint and sealant for the bolts. High Tack is a popular bolt sealer. On aluminum-block engines use anti-seize compound.

Clean & Check Decks—To make absolutely sure the head-to-block mating surfaces on both the block and the cylinder heads are clean, go over them with a solvent-wetted paper towel. Check for nicks, burrs, old gasket material or any other irregularities that might need attention.

Check that two hollow locating dowels are in each deck surface. If they are missing, they must be replaced. You'll find these dowels at your Chevy dealer, of course, or your engine machinist or a junkyard. The blunt end of the dowel goes into the block. Just tap the dowels in with your hammer, being careful not to deform the tapered end.

Install Head Gaskets—Examine the head gaskets for any cuts or nicks. If you spot any damage, take them back and exchange them. Your best gasket

insurance is the use of a high-quality item in the first place. You'll find that gasket quality does make a difference.

Make sure you know which way the gaskets fit on the heads and block. Go through a couple of "dry fits," holding the gasket against the head or block to make sure all the holes match. Gaskets are usually marked with both TOP and FRONT, so you shouldn't have any trouble figuring out how they should be positioned on the block.

Seal or Not to Seal?—Here's where the high-temperature aluminum paint or High-Tack comes in. Give each head gasket an even coat; don't slop it on. The paint will help the gaskets seal, although Chevy says you don't have to use any sealer with their gaskets. Every "pro" mechanic I talked to would never put a big Chevy together without head-gasket sealer, and it certainly can't hurt, so spray away. The pros seem to like High Tack or Copper Coat for gasket sealing.

There's one exception to painting head gaskets. If you are using the type of composition gasket that is specifically designed not to need retorquing, there's no need to paint the gasket. Fel-Pro's Permatorque head gasket is a good example. This composition gasket has a blue residue that contacts the head and block. If you painted these gaskets, the special chemicals could not work properly. If in doubt, read the instructions accompanying the gasket set. If the gaskets are to be installed dry there will be a warning to this effect.

After the paint dries so you can handle the gaskets, lay them on the block so they engage the dowels.

Install Cylinder Heads—Before you set a head on the block, make sure the engine is supported so it won't fall or roll over. If the engine is on a stand, tighten or pin it so the engine can't rotate. If you have the engine on a bench or the floor, block it up so it won't fall over. The engine must be well supported.

Have one head bolt for each head close by. Carefully set the first head on the block, making sure the dowels are firmly engaged *before* you let go. Run one bolt into the head and block so there's no chance the head can fall off. If one of these heads were to fall off the block, it could severely hurt you and the head. Before you install the head bolts, set the other head on

Torque head bolts using proper sequence and increments of torque. Deviating from this procedure could result in a blown head gasket.

Coat lifter foot with moly grease.

Lubricate lifter body with EOS or motor oil.

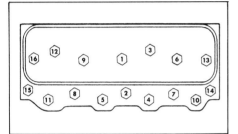

Torque head bolts in 10-ft-lb increments to 60 ft-lb using above sequence. Increase torque to 65 ft-lb—no more for short bolts—and 75 ft-lb in long bolts following the same torque sequence.

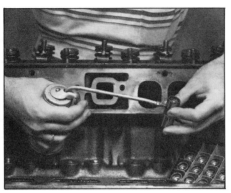

Priming lifter by forcing oil through lifter bleed hole. Lifter is primed when oil comes out top of lifter.

Insert lifters into lifter bores. A final warning: If you are reusing old lifters, they must go back to their original bores.

the block using the same procedure.

Prepare Head Bolts—Give the bolt threads a coat of sealant, then drop the bolts into their holes in the cylinder heads. You should have four short bolts along the bottom of each head. The long ones go in the remaining bolt holes. Run them in with a ratchet, following the head-torquing sequence in the illustration. Don't start at one end of the head and go to the other, or you will compress the gasket incorrectly.

After snugging the head bolts with the ratchet or speed handle, you are ready to torque them. You must tighten the head bolts in several gradual steps, not one big one. Start by torquing all the *short* bolts along the lower edge of the head in exact sequence, in 10 ft-lb increments. When you get to 60 ft-lb, final torque the short bolts 65 ft-lb and the long ones 75 ft-lb in the same sequence. Double-check bolt torque on this head when you're finished torquing the other head down, and vice versa.

After all the work you went through to get the inside of your engine clean,

you sure don't want dirt getting in through the spark-plug holes. Prevent this by installing a set of throwaway spark plugs. Use old plugs that you don't mind getting broken, because before you get the engine fully assembled you can bet one of those plugs is going to get knocked. There's no need to wrench-tighten the plugs, just run them in finger-tight.

VALVE TRAIN

Next are the lifters, pushrods, rocker arms and their balls. In addition you'll need a squirt can of oil, some moly lube or a can of EOS.

Install Lifters—Prepare the lifters by wiping them and their bores clean with a lint-free paper towel. Prime each lifter by squirting oil in the small hole in its side with an oil can; that's if they are hydraulic. Another way to do it is to place the lifter foot down in a container of oil so the small hole is submerged, then work the plunger up and down with a pushrod. This draws oil into the lifter. Mechanical-lifter engines have "solid" lifters, so there's nothing to prime.

Both types of lifters need their sides lubed with oil and their feet with moly. After lubing the lifters, drop each one in *its* bore. If you are installing your old lifters, they *must go back in the same order as they were originally installed*. New lifters can go in *any* bore.

Install Pushrods—Pushrods are similar to lifters. Old ones must go with their original rocker arms, new ones can go anywhere. The pushrods need a dab of moly on each tip before you pass them through the head and into their lifters. The top portion of each pushrod should rest in its guide.

Install Rocker Arms—You can install the rocker arms on any stud as long as that rocker arm has its original rocker ball. Lube both of the rocker-arm working tips with moly, one end where the pushrod attaches and the other where the rocker contacts the valve. Drop the rocker over its stud.

Lube the rocker ball with EOS, and drop it over the stud and into the center of the rocker arm. Finally, run an adjusting nut most of the way down the stud. Don't bother to con-

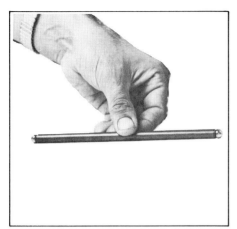

Check pushrods for straightness, regardless of whether they are new or not. Rolling them on a flat surface will uncover a bent pushrod; it will wobble. If one wobbles, don't attempt straightening it, replace it. Photo by Tom Monroe.

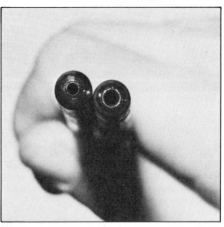

When a pushrod tip is worn out, it is obvious. Look at oil hole in pushrod tip at right. Replace complete set unless you are building a "short-timer."

Long pushrods are for the exhausts; short ones are intakes. Lube pushrod tips with moly or EOS. Insert pushrods through each head and into its lifter and guide plate. Rockers, *their* balls and nuts go on next.

Hydraulic lifters and mechanical lifters require different valve-adjustment procedures. First, lifter must be on cam-lobe base circle, regardless of lifter type. Hydraulic lifters: Tighten adjusting nut so all slop is out of valve train, but pushrod can be rotated as shown. Give adjusting nut one full turn. Mechanical lifters: Back adjusting off until rocker-to-valve clearance is 0.024 in. for intakes and 0.028 in. for exhausts. Use feeler gages to measure clearance.

tact the rocker ball with the adjusting nut yet.

Valve Adjustment—There are two ways to adjust the valves, depending on the lifter type—hydraulic or mechanical. I'll cover the more common hydraulic variety first.

Hydraulic lifters are easy to adjust because the lifter is designed to do some of the work for you. All you have to do is get the adjustment in the ballpark; the lifter will compensate for the rest. One operation you must perform when adjusting *any* valve is to make sure the lifter is on the base circle before you make any adjustments. This is so that all excess compliance—*lash or slop*—in the valve train can be adjusted out and

the valve will open fully in relation to camshaft-lobe lift. Luckily, because of the way the lobes are arranged on the camshaft, you can adjust several valves before you have to reposition crankshaft.

Begin by positioning the piston in cylinder 1—it's at the left front of the engine—on TDC of its firing stroke. Do this with a 3/4-in. socket and breaker-bar on the damper bolt. Turn the crankshaft while watching the piston, damper timing marks *and* cylinder-1 lifters. Don't bother looking at the rocker arms because they are not tight enough to matter yet.

Double-check: Both the intake and exhaust lifters should be "closed"—on the lobe base circle—at the same time. Also, cylinder-1 piston should be at the top of its bore and the damper timing mark at 0, or TDC. This is where the crankshaft must be before you adjust cylinder-1 valves. You can also adjust cylinder-2, -5 and -7 intake valves, plus cylinder-3, -4 and -8 exhaust valves.

The adjustment is done by tightening the rocker-adjusting nut. Set your 5/8-in. socket and ratchet handle on the rocker adjusting nut and operate it with your right hand. With the fingers of your left hand, feel the pushrod. Tighten the adjusting nut so the pushrod feels firm, but you can still turn it with your fingers. This occurs when the rocker arm forces the pushrod into the lifter and *begins* to compress its plunger. Note the position of your ratchet handle when this occurs, then give the adjusting nut one full turn. The valve is adjusted. Adjust all the valves in the same manner.

When you've adjusted all the valves in the first sequence, go back to the damper bolt. Turn the crank clockwise one full turn—360 crankshaft degrees—until the mark is again at 0, or TDC. Look at cylinder 6 to verify that the piston is at TDC. Cylinder 6 is the third cylinder to the rear on the right side. Its lifters should also be in the "closed" position. Now you can adjust the intake valves in cylinders 3, 4, 6 and 8. Also adjust the exhaust valves in cylinders 2, 5, 6 and 7.

Mechanical Lifters—When adjusting mechanical lifters you can use the same crankshaft-positioning sequence outlined above for hydraulic lifters. Instead of feeling the pushrod though, you will need a set of feeler gages to set clearance, or lash, as opposed to the zero-lash hydraulic lifters. This clearance is set between the rocker-arm and valve-stem tips.

This clearance ensures the valves will fully close. The actual distance used when *lashing* mechanical lifters compensates for the expansion or "growth" of valve-train components go through as they warm up. The more these components expand, the greater the lash will have to be. Hydraulic lifters require no lash because they can compensate over a limited range by filling or draining oil. Therefore you only need to be close with hydraulic lifters. Not so with mechanicals. You must be precise.

For the high-performance, street-mechanical cam, do this first *cold lash* at 0.024 in. for the intakes and 0.028 in. for the exhausts. The valves will run a little loose once the engine warms up the first time you run it, but that won't hurt anything. After retorquing the head, readjust *hot lash* to the same specifications, page 150. The valve train will quiet considerably. What you want to avoid is running the new valve train with too-*tight*—not enough—clearance. That can start valve burning right off the bat and promote rapid camshaft wear. Make sure you set valve lash properly.

WATER-PUMP INSTALLATION

Installing the water pump is next. If it hasn't been cleaned and checked yet, give it the solvent treatment. You'll need the four, 3/8-16 mounting bolts. Two of these bolts may be specially constructed to mount engine accessories. If so, they will have the normal threaded shank, then a hex head, then an extra threaded shank, or stud, on the hex head. Install these bolts in the top holes.

There are two gaskets, one for sealing each pump outlet at the block. Install the gaskets on the pump—sealer on both sides—feed the bolts through the water-pump housing, then attach the unit to the engine. Hold the pump against the block with one hand and finger-tighten the bolts with the other. Once the pump is snug against the block, torque the bolts 30 ft-lb.

Most of the time, bolting on the water pump is all you have to do with the unit. If the heater inlet and outlet nipples are badly corroded, replace them. This job is a lot easier with the pump bolted to the engine, because it will be held securely. New fittings are best purchased at a parts house. Most dealers don't stock them. Both fittings have pipe threads, the larger is 3/4 in., the smaller is 5/8 in.

INTAKE MANIFOLD

Before covering up the valve train, pour any remaining EOS over the lifters and rocker arms. You were going to put it in the oil anyway, so why not pour it where it will do the most good? Get the intake manifold out of storage along with its 16 3/8-16 bolts, the two rubber end gaskets and two sandwich-type gaskets. You'll also need both RTV and adhesive-type sealers for these gaskets, so have them ready.

Rubber Gaskets First—Begin the manifold installation by laying the rubber gaskets on the block. Each gasket has three locating dowels that

Sealer and two gaskets go on water-pump outlets. RTV or non-hardening sealer is fine here.

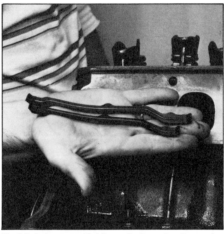

End intake-manifold gaskets: short one is for passenger cars and light trucks; long one is for tall-block 366- or 427-truck manifold. Use the two you need and discard the others.

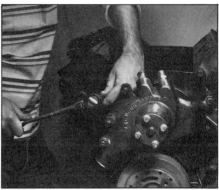

About the only complication with installing water pump is getting special bolts in the right holes. Bolts with studs projecting from their heads go in top holes; they are for mounting accessories.

Lay end intake-manifold gaskets in place and put a dab of RTV sealer on their ends. Run a bead of sealer around cylinder-head water passages and place head-to-manifold gaskets on heads. Also seal around water passages on manifold-side of gaskets.

fit into matching holes in the block. These gaskets do a fine job of sealing the block "valley" by themselves, but a light application of RTV or adhesive sealer never hurts. One problem with rubber manifold gaskets is that they tend to squeeze out of place as the manifold is tightened down. The dowels on big Chevy gaskets help prevent this, but you should realize that slippery RTV could cause the gaskets to slide out of place. I use RTV in this application, especially at the corners, because it does an excellent job of sealing.

Next, install the two sandwich-type gaskets. It is important that these gaskets seal or your engine will have coolant and "vacuum"—air—leaks. Like the end gaskets, but even more so, the "manifold gaskets" can shift

during manifold installation. To avoid this, use adhesive sealer. Apply the sealer to the head side of each gasket, place the gasket on the head and pull it off and on until the sealer is tacky. Then press the gasket onto the head.

After they are on the heads, apply sealer to the manifold side of the gaskets. Dab RTV in the four corners where the rubber and sandwich gaskets meet. Now you can set the intake manifold on the engine, being very careful to line up the manifold as you lower it onto the gaskets. If you move the manifold after setting it down you will probably dislodge the gaskets.

Installing an intake manifold is not easy. Find a helper to share the work, each of you taking an end. After the manifold is in place, inspect the gaskets as best you can to see whether

To keep gaskets from being shifted out of place, extra care is essential when laying manifold on engine. Lay manifold squarely onto engine—it's harder than it looks. So take your time. If you can enlist the help of a friend, do so.

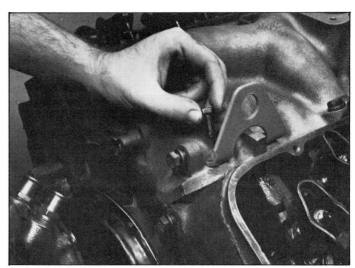

Don't forget about brackets when installing intake-manifold bolts. And if manifold is aluminum, don't forget steel washers that go under bolt heads.

Torque intake-manifold bolts 30 ft-lb using the proper sequence and increments to ensure good sealing.

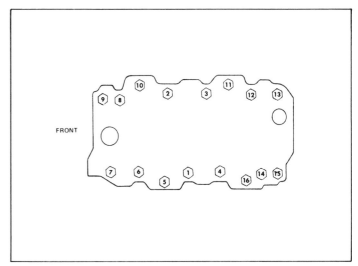

Torque intake-manifold bolts 30 ft-lb in the sequence shown.

they have shifted. Now drop the 16 bolts in place, but only after fitting any mounting or lifting hardware that installs under them. If you have an aluminum intake manifold, the bolts must have hardened-steel washers under them or they will dig into the manifold.

Following the torque sequence shown in the accompanying chart, snug the bolts with a 9/16-in. socket/ratchet. Final torque for the intake manifold bolts is 30 ft-lb, but it should be torqued in three steps, 10, 20, 30 ft-lb using the same sequence.

After you install the valve covers and exhaust manifolds, come back and check the intake-manifold bolt torque. You'll probably find the gaskets have relaxed a little and you can

turn the bolts just a little more before 30 ft-lb is reached.

Valve Covers—Each valve cover has four small cutouts in the turned-down gasket flange that runs around its perimeter. The valve-cover gasket has four triangular-shape tabs that fit into these cutouts and help hold the gasket in place. Their most important job is to keep the gasket from falling into the rocker-arm area and starting a leak. The best way to install the rocker-cover gaskets is to seal them to the rocker cover, but *leave them dry* on the head side. This allows you to remove and replace the rocker covers for valve adjustments and head torquing after the engine has been run.

Once you have the gaskets in place, install the rocker covers on the heads,

securing them with their 1/4-20 bolts and specially shaped flat washers. Don't overtighten these little bolts or you'll only bend the rocker covers. A mere 2 ft-lb is the specification, and that doesn't require much more than a touch of the wrench to the bolts.

Exhaust Manifolds—Before installing the exhaust manifolds, inspect them for cracks around their mounting flanges. If you find any, replace the manifold.

To install the exhaust manifolds you'll need the manifold gaskets and 16 3/8-16 manifold bolts. Just as a curiosity, the exhaust manifolds are different; the right one is a *log-type* manifold while the left example is a less-restrictive, freer-flowing design. Don't worry, they'll only go one way

If you use sealer on rocker-cover gaskets, put it on the rocker-cover side only. Gasket ears hold gaskets in place if sealer is not used. Don't overtighten rocker-cover bolts—a tad over 2 ft-lb is all that's necessary. Much more will distort flange.

Now is a good time to paint engine. Mask all openings such as those in intake manifold. Overlap flat surfaces with masking tape and trim it with a plastic or rubber mallet.

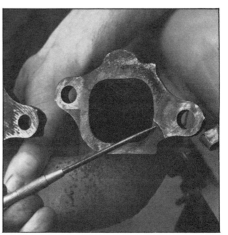

If you haven't inspected exhaust manifolds, do so now. Use one like this and you'll get a click, click sound from your engine. Comb the junkyard for a replacement manifold.

Shiny side of gasket goes against exhaust manifold. Thread old sparkplugs into each head and position manifolds for spark-plug-socket clearance before snugging bolts. Torque bolts 20 ft-lb; retorque after gasket relaxes.

Push dipstick tube all the way into oil pan. Secure tube at the top with rear cylinder-4 exhaust-manifold bolt.

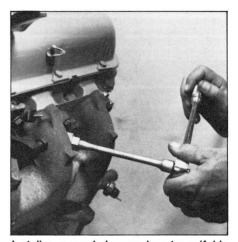

Install new sparkplugs; exhaust manifolds shield them against being broken. Use anti-seize compound on plugs going into aluminum heads. Tighten gasketed plugs 1/8 turn; gasketless plugs 1/16 turn. See text for exact torque. Don't torque cylinder-1 plug. You'll remove it later.

and make any sense, but if you think otherwise, just look at the outside of each manifold. There should be a cast-in RH on one and an LH on the other.

Use an adhesive-type sealer on the sandwich-type gaskets if you want. The shiny metal side goes against the manifold, so paste it on before applying sealer to the other side.

To install a manifold on its head, feed two bolts through the first and last bolt holes, then place the manifold against the head; run in the two bolts. It is difficult to hold the manifold without letting it drop when running in the bolts, but do the best you can. The less manifold movement, the less you'll have to fish around repositioning a slipped gasket. Install and snug the rest of the bolts, then torque them

to 20 ft-lb. No particular torquing pattern is required, but it's a good idea to tighten the bolts in steps working from the center out until 20 ft-lb is reached.

Dipstick Tube—The dipstick tube is retained by cylinder-4 exhaust-manifold bolt. Install it after you've checked that its O-ring is in place and in good condition. Push tube into the pan. Then install the manifold bolt.

SMALL PARTS

Your engine really looks like something now. All pretty and clean, it seems ready to go. But there are still quite a few pieces to install on it. Most of them are best bolted on after the engine is in the chassis—the carburetor and distributor come to

mind—but there are a few you can install now a lot more easily than after the engine is in.

Paint—Now's a good time to spruce up your engine with a coat of paint. Start by preparing the engine. Wipe it down with a rag and lacquer thinner to remove the assembly oil that ran down the sides of the block, heads, oil pan and the like. If you don't, the best engine paint in the world won't stick. Next, mask off openings and mounting surfaces, such as those for the fuel pump, oil filter, carburetor and oil-pressure sender; don't worry about the exhaust ports. Now, shoot the engine with a good grade of engine enamel; Chevy red would be a good color choice.

Sparkplugs—Now that the exhaust

manifolds are in place, you can safely install new sparkplugs. The manifolds will shield them against breakage. I'm not saying it's impossible to break one during engine installation, but the chances are slight. So, for once in your life, enjoy a cold-engine, walk-right-up-and-change-'em sparkplug change. Remember to check the plugs for the proper gap and install them at the proper torque.

Earlier engines with cast-iron heads use 13/16-in. plugs that should be twisted to 25 ft-lb. After 1970, all cast-iron heads have the smaller 5/8-in. "peanut plug" that is torqued 15 ft-lb. *All* aluminum heads use the larger, 13/16-in. plug, but is installed at the lower 15-ft-lb torque. You should coat all plugs, but especially those going into aluminum heads, with anti-seize compound.

Be careful when applying anti-seize compound to the plug threads. If it gets on the plug tip, it will cause fouling. Keep anti-seize compound off of the first plug thread—the one closest to the plug tip. And, *don't put anti-seize on the plug threads in the head.* Otherwise, the plug will push the compound ahead of the plug, causing it to foul.

Install all the plugs to their final torque except for cylinder 1. You'll need to remove it to check cylinder-1 piston position later. For now just run cylinder-1 plug in finger-tight.

Oil-Pressure Sending Unit—If you are careful you can install this sending unit and get it into the car safely. Otherwise you'll have to crawl underneath to thread it in above the oil filter. It's a toss-up, look at your chassis and decide. Tight Corvette and passenger-car engine compartments could mean a broken sender, while a truck chassis poses few problems. Before threading in the sending unit, coat the threads with a non-hardening sealer.

Fuel Pump—On the passenger-side of the engine, install the fuel pump and pump-to-carburetor fuel line. Slide the fuel-pump pushrod into its bore after dabbing each of its ends with moly grease. The pushrod slides through a hole just below the fuel-pump opening. You probably have the plug that fills this hole finger-tight, so remove the plug, slide the pushrod in place. Hold the pushrod all the way up its bore until you have the plug in place and tightened down.

Apply sealer to both sides of the

If oil-pressure sender can be installed with engine in place, install it later. If bumped as the engine is being manuevered into place, sender will break off. Regardless, coat sender threads with non-hardening sealer.

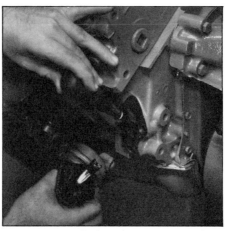

...hold pushrod with a finger or screwdriver so you can get fuel pump in place. Or, if you're assembling engine on a stand, turn engine on its left side and let gravity work for you. When you have pump oriented correctly—diaphragm down—pull your finger or screwdriver out and quickly insert pump.

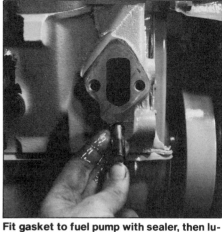

Fit gasket to fuel pump with sealer, then lubricate fuel-pump pushrod. EOS or moly works well. Slide pushrod into position, then...

Run in two bolts to hold pump. If pump won't mate against block without a lot of pushing, rotate crank one revolution. This will rotate fuel-pump eccentric and let pushrod retract.

fuel-pump gasket, stick it to the pump and fit the pump to the block. The pushrod may have slid down its bore, so you'll have to hold it in place while installing the pump. Use your 9/16-in. socket/ratchet combination to tighten the two fuel-pump bolts. If you find yourself fighting the fuel-pump return spring, turn the crankshaft about one revolution to bring the camshaft eccentric to its low point.

Fuel Filter—Before you slip on the flexible fuel line from the fuel-tank-to-pump line—which is done after engine installation, so you have a little time yet—consider adding a fuel filter ahead of carburetor. Any of the paper-element filters in clear plastic housings will ensure that clean fuel will flow through the pump and to the

carburetor. Gasoline is often dirty because the underground tanks it is stored in are pumped-dry, allowing all the bilge-drainings to end up in your tank. Aftermarket fuel filters are very cheap and do an excellent job of screening out harmful particulates.

To install one, cut the tank-to-pump fuel line several inches after the fuel pump. Point the arrow on the filter body towards the pump, then clamp both ends of the line to the filter. Finish by clamping the rubber fuel line to the fuel-pump inlet. The inlet fitting is marked IN.

Engine Mounts—Attach both engine mounts to the block. Three 3/8-16 bolts secure each mount.

I recommend installing new engine mounts. The big-block is heavy and

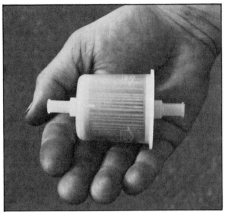

Inline fuel filter installed between pump and carburetor filters out dirt before it gets to carburetor. Clear filter lets you see any water or debris.

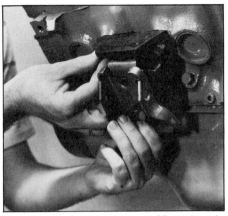

New engine mounts are a good investment. Position mount so the wide metal section is at bottom.

One of the final parts to go on before engine installation: the crank pulley. Install bolts and tighten them evenly. Check pulley to make sure it is square with damper.

the hostile environment limits engine-mount life. A failed mount—usually it's a separated metal-to-rubber bond—will allow the engine to move around so much that the throttle and transmission linkages could bind. Unless you installed new mounts a few thousand miles ago, install new ones now. This will be considerably easier than after the engine is back in its compartment.

Crank Pulleys—Bolt the crankshaft accessory-drive pulleys to the damper now rather than after the engine is in the chassis; it'll be considerably easier. Three 5/16-18 bolts secure the pulleys to the damper. A dab of Loctite on their threads will ensure that they won't come out when you don't want them to.

Thermostat & Housing—This is a good time to test your old thermostat, particularly if your engine had overheating problems. Also, if the thermostat is new, you should test it for sure. Pro mechanics know one out of every 11 *new* thermostats are bad right out of the box, so testing a new one is not a waste of time.

Test a thermostat by hanging it by a wire or string in a pot of water on the kitchen stove. Don't let the thermostat touch the bottom or sides of the pan. You also need a cooking thermometer to check water temperature. You'll have to ask you-know-who where she keeps it. Turn the stove all the way up while monitoring the thermostat. When the thermostat begins to open, check the thermometer for the water temperature. If the thermostat hasn't cracked open by 195°, or isn't distinctly open by the time the water boils at 212°—return it

for another. Don't forget to test the replacement thermostat.

When you finally check out a working thermostat, lay it aside until the engine is in the chassis. Installing the thermostat after the engine is filled with water avoids air locks that could cause your new engine to overheat on initial startup.

Intake Manifold—You may have stripped the intake manifold for cleaning, so now you'll have to "dress" it. First install the separate choke heat sensor. Tighten it in place with its single screw. If you can't find where it goes, look under the right side of the carburetor-mounting pad.

The water-temperature sender can be replaced now if you don't think you'll break it while installing the engine. I prefer to leave the senders out until after the engine is in the chassis. If you put it in now, seal its threads with a dab of Permatex No. 2 or another non-hardening sealer. Early senders go into the intake manifold; post-'68 senders go in the left cylinder head.

Also on the leave-off list are the carburetor and distributor. Both are easily damaged during installation, and are almost as easy to install after the engine is in place.

If you haven't put new water fittings in the intake manifold and water pump, give the old ones an inspection and replace as needed. Install the water-pump bypass hose. That's the small, 90° hose that runs between the water pump and intake manifold. Also seal and install the vacuum tee in the rear, upper intake-manifold runner. If you have engine-lift lugs and forgot them during manifold

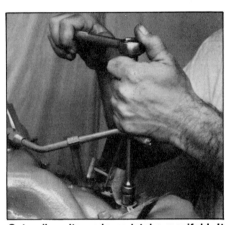

Set coil on its pads on intake manifold. It should lean to rear of engine.

installation, make sure they go on now.

For all of you with conventional ignition systems, the coil mounts at the intake-manifold rear. The coil-mounting bracket gives the coil a rearward lean, so if it looks like the coil just won't fit, turn it so it leans to the rear of the engine with the bracket feet flat against the manifold. Two bolts secure the coil.

That's it for engine assembly.

8 Engine Installation

Installing and starting your engine is the last major operation of the rebuild. Everything in your parts pile should be used up or replaced when you are done. So, if there is any other work you want done, such as rebuilding the radiator or replacing the starter, do it now. At the least, the parts should be clean and neatly arranged.

One job that will save a lot of frustration is charging the battery. It sure is maddening to try to start a new engine, only to find that the battery that has been sitting under the bench for a few weeks is so dead the dome light won't even come on. If you don't have a battery charger, take the battery to a service station and have it charged. Or ask a friend for the loan of his charger.

Clean Engine Compartment— Another often-neglected job is cleaning the engine compartment. It would be a shame to drop your new engine into a compartment with 150,000 miles of accumulated crud. A clean engine compartment makes engine installation easier and safer besides. Any sort of degreaser works well, but canned engine cleaners seem to work best. Don't approach this cleaning job haphazardly or you could harm the car's mechanicals, electrical system or cosmetics.

Take the time to mask parts that should not be wetted. Also, engine cleaners contain strong detergents and will harm exterior paint, so be careful not to splash it carelessly on the vehicle exterior. If you steam-cleaned or pressure-washed the engine and its compartment before pulling the engine, tidying up the compartment should be easy.

Another way to do your cleaning is at the local car wash. It's a lot of work to tow the car to and from the car wash, but at least you leave the mess there and have the use of high-pressure hot water. You can also clean loose bolts, brackets and other hardware at the car wash without much trouble. Don't forget to clean the underside of the hood.

Take Your Time—Before getting into the mechanics of installing the engine, I must caution against rushing the job. It's only natural to be overeager when installing a new engine. It is the last part of a long and involved job, so wanting to hurry and finish the work and see the results is understandable.

Work as quickly as you wish, but remember, do each job *thoroughly* and *carefully*. It doesn't make any sense to spend a lot of time and money pulling, tearing-down, inspecting, and assembling an engine only to do a hurry-up, haphazard job of installing it. You may end up wasting your time and money if you do. Work patiently. If you find yourself in too big a hurry, force yourself to quit for the day and start again tomorrow.

TRANSMISSION FRONT SEAL

While the engine is out, you can easily replace the front transmission seal. This seal fits around the transmission input shaft on both automatic and manual transmissions. If there is any sign of leakage inside the bellhousing that did not come from the engine, you must change the front seal. If it's an automatic, I recommend

that you replace the seal regardless. The reason is simple. If you don't, chances are the seal will develop a leak soon after you have your vehicle back in operation.

Automatic Transmission Seal—To change the seal, start by measuring the position of the converter with respect to the transmission. Lay a straight edge across the front of the transmission and measure back to a converter-to-flexplate mounting pad. Push the converter back toward the transmission to make sure it's fully installed so you get a good measurement. Record this figure; you'll need it later when installing the converter.

Remove the torque converter. To do this, simply pull the converter straight off of its shaft. Be careful, the converter is heavy—about as heavy as one cast-iron head. In addition to the weight of the converter itself, it should be no less than half-full of automatic-transmission fluid (ATF). And as you pull the converter off, tip it down at the front to prevent the ATF from pouring out.

After you have the converter clear

435-HP 427 at home in its engine compartment. A clean engine compartment will complement your newly rebuilt engine.

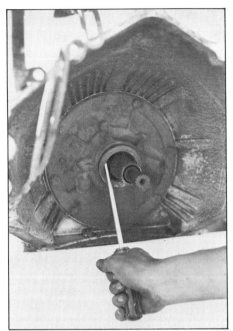

Replace front transmission seal if yours is an automatic. Otherwise, expect a transmission leak later.

of the engine compartment, tilt it over a bucket or drain pan and pour out the ATF. When you have the seal installed, replace the torque converter dry. You can fill the converter more accurately and with less mess later through the dipstick tube.

With the converter out of the way, the seal will be visible. Pry the old seal out with a screwdriver, taking care not to gouge the stator shaft or aluminum transmission housing. Work around the seal; don't pry in one spot only.

Clean the seal bore with a paper towel and lacquer thinner to remove any oil in the seal cavity. Run a bead of sealer around the seal OD before installing it. Lube the seal lip with some fresh ATF.

Install Seal—Just like other metal-housed seals, the front transmission seal must be started and driven in squarely. I doubt you have a socket deep or large enough to fit over the transmission input shaft and stator support. If you don't, a pipe or steel tube will work. If you strike out here, don't worry. You can install the front seal with a hammer.

Hold the seal squarely over its bore,

lip pointing toward the transmission. Begin working it in with light taps around its periphery. Start with taps opposite each other: at 3 o'clock, at 9 o'clock, 12 o'clock and so forth. After the seal is well started, go to an around-the-clock tapping sequence. When the seal bottoms you'll feel it, so don't overdo the tapping. Wipe excess sealer from the seal and transmission and, while you're at it, clean the entire inside of the bellhousing.

While you still have access to them, note the ring of bolts holding the pump cover to the front of the transmission. Torque them 18 ft-lb to tighten any that may have loosened.

Install Converter—The converter is installed by slipping it onto the transmission shaft and into engagement with the shaft, stator support and pump. A couple of tricks will help the installation.

First, inspect the torque-converter seal surface for any roughness or varnish. If the surface is rough, give it the same 400-grit-paper treatment you gave the crankshaft journals. If it's just coated with varnish, clean it with solvent or lacquer thinner.

Next apply some ATF to the converter-seal surface to lubricate the new seal. Grasp the converter firmly, then lift it up to the transmission input shaft. Don't let the converter hang from the shaft while you change your grip. This may damage the seal or the shaft. You should be able to install the converter with one steady motion.

Move the converter back into engagement with the shaft, stator support and pump—three separate engagements. You should have little trouble if you know what to expect. The converter will stop moving on the shaft as if it were fully installed, but your better judgement says it isn't. And your better judgement is probably right. Push back, rotate and lift up on the converter nose and the converter should slide on into its three engagements, the last one being the pump. Each should be distinct. Check the position of the converter using the converter-to-transmission figure you recorded earlier. If it's not on all the

way, keep turning and pushing until the converter is fully engaged.

The last step in the installation is to rotate the converter so the marked converter-to-flexplate connection is at the bottom of the bellhousing.

Manual Transmission—Unlike automatic transmissions, manual gearboxes rarely leak. If your standard-shift box has never leaked and isn't leaking now, don't bother the front seal. You'll only be disturbing a good thing. However, if the front seal has been leaking, as evidenced by heavy, sweet-smelling grease in the bellhousing, change it now.

To replace the front seal, unbolt the input-shaft-bearing retainer. It's the round piece circling the input shaft that the clutch-release bearing slides back and forth on.

Start by removing the clutch-release fork and bearing. Push the fork inward to unclip it from its pivot. Remove the four bolts from the bearing retainer and remove the retainer.

Turn the retainer nose down and you'll see the seal. To remove it from the retainer, you'll need a long punch or a socket-wrench extension. Insert the punch in the bearing-retainer nose and butt it against the backside of the seal. Work around the seal and tap it out of its bore.

After wiping the seal bore clean, coat the seal OD with RTV. Place the new seal squarely over the bore with its lip pointing up and the nose of the bearing retainer pointing down. With a socket that fits squarely against the seal metal edge, drive it into place until it bottoms. Lube the seal lip with fresh oil and fit a fresh gasket to the retainer, using RTV to seal it. Install the retainer and tighten the bolts.

ENGINE PREPARATION & INSTALLATION

With the chassis work out of the way, turn your attention to the engine and the parts that bolt to it.

Clutch & Flywheel—In case of a manual transmission, you must inspect the clutch before installing the new engine. Now is the logical time to do such a job—while the engine is out.

Big-blocks, with their high torque

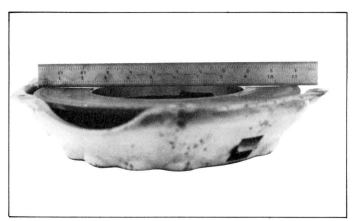

Warped pressure plate is indicated by gap between straightedge and friction surface. Photo by Tom Monroe.

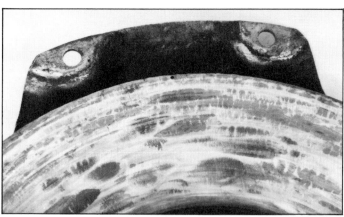

Heat checks and cracks indicate pressure plate was severely overheated. Such a pressure plate is dangerous. Replace it. Photo by Tom Monroe.

output, have a tendency to wear out clutches, particularly those in sporty-type cars, such as Corvettes and Chevelles. If you have one of these cars, it's not for puttering around the neighborhood as if you were driving a six-cylinder-Powerglide combination. So, if you own a stick-shift big-block, and especially if you are not the original owner, count on replacing the clutch. There is one exception; you may be one of the lucky few to have a double-disc clutch. In this case, it probably won't be worn out. Although a double-disc will rarely wear out, check it anyway.

Three basic flywheels are used on big-blocks. The first is the 14-in. with a 168-tooth ring gear. It is used with the 10-in. double-disc clutch and 11-in. single-disc clutch. The second is 12.75-in., 153-tooth ring-gear flywheel. Only 10- or 10.5-in. single-disc clutches are used with this flywheel.

The third is almost the same as the 12.75-in. flywheel just mentioned, but is counterbalanced for use with the externally balanced 454. Other big-block flywheels are *zero balanced*—no counterweighting.

Clutch & Flywheel Interchange—If you are interested in moving up to the dual-disc clutch or the 11-in. single-plate version, you will have to have the 14-in. flywheel. If you already have a 14-in. flywheel, these two clutches will bolt right on. All you have to do is pay the steep price for the double-disc clutch. If you have a 12.75-in. flywheel you'll need the 14-in. flywheel, a larger bellhousing, 3899621, and starter-motor housing 1969309. By the time you add all the parts up you would probably pay less

for a new Turbo Hydra-matic 400 automatic transmission—if you want an automatic.

Inspect Clutch—No matter which clutch you have, the wear scenario is the same. Friction between the disc and pressure plate wears down the disc and heats the pressure plate. Heat causes hard spots to form on the pressure plate, which show up as blue-black spots. Heat also warps the pressure plate and fatigues the springs. On the other side of the disc, the flywheel gets a similar treatment. But because the flywheel has considerably more mass than the pressure plate, it is able to take more punishment before it is damaged.

Pressure-Plate Inspection—All Chevy clutches are of the *Belleville* design. This design is characterized by the lack of coil springs. Rather, it uses a single, flat disc-type spring. The part of the spring you see through the center of the pressure-plate cover has many radial cuts that form the clutch *fingers*. These fingers extend toward the center of the pressure plate and rearward at a slight angle. When sighting across the spring fingers, they should appear to be at the same height. If they aren't, the spring could be cracked.

Inspect the area where the fingers meet the uncut portion of the spring for cracks. If you find any, the spring must be replaced. If the spring is OK, turn the pressure plate face up and lay a straight-edge against the pressure ring. If it is warped, you'll see a definite concave shape beneath the straight-edge. Also, inspect the pressure plate for hard spots and *heat checks,* or cracks. Look for grooves cut into the pressure plate by clutch-

disc rivets. If the disc was badly worn so its rivets were exposed, the pressure plate and flywheel will be grooved.

You have a choice when it comes to servicing the pressure plate. The first method is resurfacing your old pressure plate if it's warped or grooved. This is done in a machine shop by grinding metal from the pressure-plate working surface. If it has surface cracks, you have no choice. It must be replaced. This brings us to the second choice; replacing the pressure plate.

Replacing a pressure plate is preferable to resurfacing an old one for several reasons. First, the pressure plate will be full thickness and thus be better equipped to absorb heat. Second, the spring will be new and not fatigued like your old one, so you will get maximum capacity from your clutch. Third, other internal parts will be new, not partially worn. Fourth, the rebuilt engine you are about to install has more power than the tired engine you pulled out a few weeks ago. This means the clutch must handle more torque. Finally, there are the price and convenience considerations. It costs money to surface the pressure plate and it won't cost all that much more to buy a new one. And, it is considerably easier to replace the pressure plate now rather than after the engine is installed.

Disc Inspection—The major indication of disc condition is the thickness of its facings. The thinner they are, the less life the disc has. Real problems begin when the disc is worn so badly that the rivets stand flush with or above the surface of the friction material. Then both pressure plate and flywheel will be grooved from rivet contact. You will have already discovered this from inspecting the pressure plate. The other disc-killer is oil. If your disc is oil-soaked, throw it away and start with a new one.

To measure disc thickness accurately, you'll have to compress it first. Use a vise, C-clamp or Vise-Grips to compress the wave large-diameter washer, or *marcel,* that separates the two friction surfaces. Remember: As you handle the disc, don't get oil or grease on it. And, clamp it only enough to flatten the marcel. It's best to fit two small pieces of clean plywood between the clamping jaws before compressing the disc.

Once the marcel is compressed flat, measure disc overall thickness with a micrometer or vernier caliper. A new disc measures 0.330 in. When the disc is worn so rivets are flush with the friction surfaces, it will measure 0.280 in. Anything approaching the 0.280-in. figure means a new disc. It all depends on how long you want the clutch to last before you must pull the engine or transmission to service it. If the disc is half worn or more, don't give it a second thought. Replace it.

Flywheel Inspection—Inspect a flywheel just as you did the pressure plate. One problem you don't have to be concerned about with the flywheel is warpage. Its size can absorb and dissipate more heat than the disc can dish out. However, look for grooving from rivets and blue-black hot spots.

Heat checking is also a problem with flywheels. Normal heat checking can be cured by light resurfacing, but cracks radiating outward from the center mean the flywheel *must* be replaced. These cracks *cannot* be removed by resurfacing. Such a flywheel is dangerous and could result in the flywheel exploding at high rpm, causing serious damage to the vehicle and possibly injuring passengers and anyone nearby.

If you have your flywheel resurfaced, have it done with a grinder, not on a lathe. The lathe cutting tool will literally jump over the hot spots, which are harder than the parent material, resulting in a bumpy flywheel surface. Grinding makes the flywheel perfectly flat the way you want it.

If your flywheel is in good condition—no cracks, grooves or hot spots—freshen its working surface with 400-grit paper. This will remove

Compress clutch disc just enough to measure its overall thickness. Minimum thickness depends on how long you want the clutch to last.

We all make mistakes. Bolts are being removed so flexplate can be turned around and installed properly. Converter mounting bosses must face converter, not engine. Rotate crank so marked flexplate hole is at bottom.

bonding resin that was transferred from the disc to the flywheel. After sanding the flywheel, wipe it with a non-petroleum solvent such as lacquer thinner or alcohol to remove any residual oil. While you have the materials out, polish and wipe off the pressure plate and disc.

Clutch-Release Bearing—There are two ways of looking at what to do with the release bearing, or *throwout bearing* as it's sometimes called. You could take the approach that every time the clutch needs servicing, the release bearing should be replaced. Or, you could simply change it now because the engine is out and it is easily done.

Either way you look at it, you'll be wise to change the bearing now because of the labor involved in replacing this fairly inexpensive part later.

The bearing is held to the release fork by two spring clips. It takes a little juggling to remove the bearing from the fork, then a good push to re-engage the new bearing. While you have the bearing off, give its bore—which slides on the transmission-bearing-retainer nose—a *light* coat of moly grease. Don't overdo it or the grease will be flung into the clutch and cause it to grab and chatter. Put another dab of grease on the release-fork pivot and at its bearing contact points.

Flywheel/Flexplate Installation—Check the crankshaft flywheel flange for nicks or other imperfections that could cause the flywheel or flexplate to sit slightly skewed. Smooth any high spots with a file. Be careful not to

make any new gouges in the process. A flywheel must be square with the crank—no wobble as it rotates—or it will have bad effects on transmission internals, the rear-main bearing and crank seal.

Position the flywheel or flexplate against the crankshaft so it engages the dowel in the crank. If there is no dowel, line up the crank and flywheel/flexplate dowel holes.

In addition to the dowel or its holes, you should have your own set of marks on the crank and flywheel to go by, those that you made during engine disassembly. Use them to double-check that this all-important crank-to-flywheel/flexplate relationship is correct. While getting the flywheel/flexplate off a space or two *may* not cause an imbalance with an internally balanced big-block, getting a 454 flywheel/flexplate mismatched would shake the engine right off its mounts.

If you're fitting a flexplate, make sure the three torque-converter mounting pads protrude *away from the engine.* With the flexplate or flywheel on the dowel pin, run in the six bolts with the serrated-edge washers. A dab of Loctite on the bolt threads will ensure that they stay put. Tighten the bolts down in a star pattern and torque them 65 ft-lb.

Install Clutch—Fitting the clutch to the flywheel can be a handful. Although, not a hard job, it can be awkward. Besides the pressure plate, the disc and six bolts, you'll also need an alignment tool. Such a tool must be used to align the disc to the crank-

Using an old transmission input shaft to align clutch disc. Alignment tool can be removed after pressure-plate bolts are snug. Photo by Tom Monroe.

Fill oil filter before installing it. The lubrication system will pressurize much quicker.

Exhaust-system "doughnuts" seal exhaust manifolds to header pipes. Remove sleeves, discard doughnuts and install new ones. You'll find them in the gasket set.

shaft pilot bushing so the input shaft will slide straight through during engine installation.

Although I've lined up a clutch disc by eye in an emergency and had it work, I don't recommend that you do it. Mating an engine and transmission is tough enough, even with a perfectly aligned disc. Doing it by eye is asking for trouble.

If you have a Chevy-transmission input shaft lying around the garage, it will do very well as an alignment tool. If you don't have one or can't find an input shaft, the local parts store has various alignment tools. They range from a cheapy wooden one to K-D's universal clutch-alignment tool. For your purposes, the wooden one will work fine.

Start the clutch installation by inspecting the disc. Wipe it down with lacquer thinner to make sure it's clean. One side must face the flywheel and the other side the pressure plate. If the flywheel side doesn't have **FLYWHEEL SIDE** stamped on it, go by the hub assembly. The hub center, its springs and retaining plate project from one side of the disc. This side installs away from the engine, toward the pressure plate and transmission.

Lay the disc on the pressure plate, hub-side down. While holding the pressure plate/disc combination together, lift the unit to the flywheel. Use one hand to hold the pressure plate and disc against the flywheel. With the other hand, insert the alignment tool through the center of the disc and into the pilot bearing.

The alignment tool should support itself and the disc while you start the pressure plate hold-down bolts. Get one or two started so the pressure plate is supported, Loctite the remain-

ing bolts. Remove the first one or two you installed and Loctite them also. Run the bolts in finger-tight. Lift up on the end of the alignment tool to center the disc, then *tighten the pressure-plate bolts gradually;* no more than two turns at a time. Otherwise you may bend the pressure plate cover, or *hat.*

Check that the alignment tool slides in and out of the disc and pilot bearing without hanging up. If it hangs up, loosen the pressure plate and try centering the disc again. If, after retightening the bolts, the alignment tool moves in and out freely, remove it and torque the pressure-plate bolts 35 ft-lb. You're now finished at the rear of the engine.

Oil Filter—Installing a standard oil filter now won't pose any problem during engine installation, but watch those extra-long filters. If you used a longer-than-stock filter on your engine before, one will be OK now. If you didn't, wait until after the engine is installed to make sure it will fit.

Before installing the filter, fill it with oil. This will quicken oil-pressure buildup during engine startup.

Exhaust Seals—A couple of items easily forgotten until too late are the two fiber *cookies,* or *doughnuts,* that fit in sleeves on the ends of the exhaust pipes. Use your pliers to extract the sleeve from each pipe. The old doughnut will probably still be wrapped around the sleeve. Find new doughnuts in your gasket set. Install them and you're ready to go.

Position Crankshaft—Engine installation is getting closer with every step. And positioning the crank on all

automatic-transmission cars is the only step left before you can drop the engine in place.

Turn the crank so the paint mark you made on the flexplate is at the bottom, or 6-o'clock position. Double-check the torque convertor; it should also be at 6 o'clock. Manual-shift engines can be set to 10° BTDC on cylinder-1 firing stroke. Use the damper timing mark as reference. Do the same with an automatic after the torque converter is bolted up.

After positioning the crank, attach the lifting chain diagonally across the engine. Use the lifting hooks if your engine has them. Or, use a bolt with a large, flat washer in an engine-accessory mounting hole at the front of one head and at the rear of the other. As a reminder, don't make the lifting chain overly long or your lifting hoist will run out of travel. Now that the chain is attached, you are ready for the big lift.

ENGINE INSTALLATION

Raise the engine off the floor and adjust the angle at which it hangs. The engine should be level from side to side and slightly lower at the rear than at the front.

If you don't have helpers, recruit some now. Although you may be able to install an engine by yourself, when they are as big and heavy as a big-block Chevy, I don't recommend it. It is too difficult and dangerous to be handling this brute of an engine alone. Your helper/s don't have to be expert mechanics, either. My wife and I have installed countless engines together, and her help in watching for

With a manual transmission, set cylinder 1 on 10° BTDC. This will help speed distributor installation later.

You'll smile like Mike when you get this far with your engine rebuild. Slight downward angle at rear of engine helps with mating engine to transmission.

obstructions, steadying the engine and passing tools has always been invaluable.

Roll your engine over the car or the car under the engine, whatever the case may be. If you have a manual transmission, place it in gear. Position the floor jack under the transmission—auto or manual—and raise it slightly above its normally installed position. Remove support wires.

Lower Engine—Inspect the engine compartment for loose wires, ground-straps and the like. They must all be out of the way.

Carefully lower the engine into the engine compartment while watching for hang-ups. If the engine seems to stop going down or tilting one way or the other, *stop.* Lift the engine an inch or so and investigate.

If you continue to lower the engine regardless, you could end up with a severely bent, broken or damaged part. If you're working on a Corvette or with a vehicle equipped with a single exhaust system, you must line up the exhaust pipes with the exhaust manifolds.

Manual Transmission—With an automatic transmission, the engine can be lowered straight down, almost in line with its installed position. With a manual transmission, the engine must be lowered farther forward to clear the transmission input shaft. The clutch disc is then centered on the input shaft, then the engine is moved to the rear until the input-shaft nose contacts the center of the clutch disc.

If the clutch disc hangs up on the input shaft, rock the engine back and forth, around the crankshaft axis, in an effort to line up the splines. Make sure the block and bellhousing mating surfaces are parallel when doing this. The engine should slide into engagement with the transmission.

You may have trouble doing this because the big-block is so heavy and your engine compartment is so small. If so, there is another cure. You'll need another jack to raise one of the rear tires. With the transmission in high gear, turn the tire while a helper pushes the engine back. The shaft should rotate into engagement with the clutch disc.

If the shaft and clutch splines engage, but the engine stops short of the transmission, double-check that the engine and transmission mating faces are parallel. If they aren't, the input shaft may either be binding in the disc or its nose is not centered in the pilot bearing.

To square up the engine with the bellhousing, stand to one side and sight between the engine and the bellhousing. The gap at the top and bottom should be the same. If not, line them up by varying the height of the transmission with the floorjack, or changing the angle of the engine on its hoist. The two should line up and slide into engagement.

You'll probably have to rotate the engine to one side or the other to get the dowels at the rear face of the block to engage the companion holes in the bellhousing. When they line up, the

As soon as engine and transmission mate, start a bellhousing bolt in on each side to keep them together.

engine and transmission should snap together. Thread in two bolts right away so the engine and transmission will remain engaged until you can install the remaining bolts. You'll find the upper two easier to install from the top, the four in the corners can be reached better from below. Install them loosely, one on each side.

Automatic Transmission—Mating an engine to an automatic transmission is relatively easy. As I said earlier, you can lower the engine close to its installed position, and you don't have to worry about mating splines or shafts. The flexplate and torque-converter bolt-holes must line up, so if you have to do any fine adjustments, you can rotate the torque converter.

Run in the two or three vertical starter-motor bolts; don't forget support at front. Connect wiring.

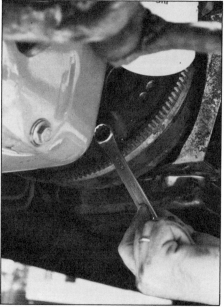

Match converter to flexplate and install bolts. Turn crank with starter motor or breaker bar and 3/4-in. socket on damper bolt. Install engine-mount bolts next.

Install flywheel/flexplate cover with a bolt in each corner.

When the trans and engine line up and the converter nose is centered in the end of the crank, move the engine into engagement with the transmission. If something hangs up, check the dowel pins. Also check that the engine and transmission faces are parallel.

As soon as the engine and bellhousing are flush against each other, run a couple of bolts all of the way in, one on each side. They will keep the engine and transmission from separating. Start the remaining bellhousing bolts and pull the engine and transmission together evenly.

Once the engine and transmission are together, you can run all the bolts in and torque them 30 ft-lb. If access to the bolts is a problem, start several bolts, snug them with a ratchet and then lower the engine onto its mounts. The remaining bellhousing bolts can then be installed.

Engine Mounts—Gently lower the engine onto its mounts. This is best done with one person looking in to line up the mounts from the underside of the vehicle. Line up the exhaust at the same time. You may have to do a little side-to-side jostling to get the engine-side of the mounts over the pad on the frame crossmember.

Feed the through bolt into each mount and install the lockwasher and nut. Once the engine mounts are secured, disconnect and remove the hoist. Remove the lifting chain. If you haven't finished installing the bellhousing bolts, do it now.

Starter Motor—Next is the starter motor. You have to get underneath to do this. Jack up the vehicle and place jack stands under the frame. Check them for steadiness. Above all, *don't trust a jack alone to support a vehicle!*

The warning I gave when removing the starter goes again—keep your head from under the starter. You sure don't want it to fall on you. With a bolt in one hand and the starter in the other, hold the starter up to the block and start the bolt. Continue to support the starter while running in the rear bolts. Torque them 25—35 ft-lb.

Bolt the support to the front of the starter and to the block. Move on to the electrical connections, if you disconnected them earlier. There are three lugs on the solenoid, a large one in the center flanked by two smaller ones.

The wiring harness that goes to the starter has three wires—small, medium and large. Fit the large red wire from the harness, along with the positive battery cable, to the large center lug. To the smaller lug, farthest from the engine block, fit the medium, black wire from the wiring harness. A manual-transmission vehicle will not have the small wire; it is from the neutral-start safety switch. If you're working on a Corvette, replace the protective sheet-metal cover over the starter and solenoid.

Flexplate to Torque Converter—If you have an automatic transmission, bolt the torque converter and flexplate together. The match-marked flexplate and converter bolt holes you positioned at the bottom should be in plain view from below. You'll have to

rotate the engine to get at the other two. Line up these holes by rotating the converter. Slip a bolt through from the flexplate side. Then run the nut on from the torque converter side and torque it 45 ft-lb.

To bring the other bolt holes in view, you can use a socket and ratchet or breaker bar on the damper bolt to turn the crankshaft. You can also use the starter after you install and connect the alternator and battery; either way is OK. Just be sure that if you use the starter method, the person at the switch keeps his hand off it until you instruct him otherwise.

When you've finished installing the remaining two flexplate-to-converter bolts, install the sheet-metal cover to the bottom of the bellhousing. This applies to manual transmissions as well.

Exhaust Manifolds—You'll need a long extension between the socket and ratchet for tightening the exhaust-manifold nuts. Mate the exhaust pipes with the manifolds. This could take a little effort, but they should go on. With a Corvette, replace the two studs you removed when disconnecting the exhaust—or use cap-screws. Slide the thick iron collar up the exhaust pipe and over the manifold studs. On the passenger side, the heat riser must also go over the studs.

Before you install the nuts, put a dab of anti-seize on the studs. The

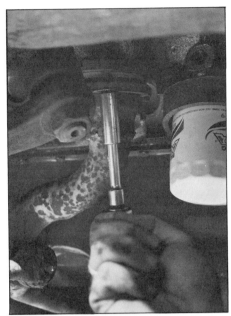

Connect exhaust system. If you have a Vette, bolts rather than studs and nuts make this job easier. Use anti-seize compound on bolt threads.

Installing oil-pressure sender after engine is in place avoids the possibility of breaking it off.

Slip on fuel hoses and install clamps. If hoses are more than two years old or are cracked, replace them.

Fill crankcase with oil: add four quarts for passenger cars and five quarts for Corvettes and pickups. Check manual for a big truck. Recheck oil level after initial run-in and add oil as necessary.

nuts will come off easier if there's a next time. Run the nuts up *evenly* with your ratchet-and-extension combination.

If you have an early fuel-evaporation (EFE) heat riser, reconnect the actuator rod to the heat-riser lever by pressing on the EFE-system rod.

If you didn't install the oil-pressure sender before dropping the engine in, now is the time to do it. Don't forget to seal the threads with non-hardening sealant.

Fuel Line—It's time to come out from under your Chevy and take a break. An easy job that needs doing is reconnecting the gas-tank-to-pump fuel line to the pump. Slip the line on the pump nipple and secure the connection with a clamp. If you haven't installed an inline filter between the pump and carburetor, now is a good time to do it.

Engine Oil—Because you already have a quart of oil in the filter, put in one less quart than normal. In passenger cars and trucks, pour in four quarts; Corvette pans are a little larger, they get five quarts. Use only the oil you are going to run in the engine. Use an *SE-* or *SF*-rated oil, certainly no less than SE. After the engine has been run in, check the oil level and top it off if necessary.

Carburetor—It makes no difference whether you have a Holley or Roches-

ter carburetor when it comes to installation. Both are held on by four bolt or nut-and-stud combinations. Check that the gasket supplied with your gasket kit is the correct one and install it dry—no sealer. *Do not over-tighten the carburetor hold-down bolts or nuts.* Gradually snug them down in a criss-cross pattern to help equalize the loads. Overtightening these bolts or nuts can cause big problems. Maximum torque is 15 ft-lb; exceeding this figure could bend or break the carburetor base.

Choke—If your car is Holley-equipped, it may have a manual choke. Installation is a simple matter of passing the cable and its conduit through the choke-cable support, then threading the inner cable through the choke arm. With the cable knob in and the choke plate wide open, tighten the clamp screws at each attachment point.

If you have an early Rochester Quadrajet or later Holley, you'll need to reconnect the choke rod. Attach this rod to the choke linkage on the passenger-side of the carburetor. Secure it with a small bent-wire clip, which should still be on the choke-control rod.

Late engines, and a few early big-blocks, used carburetors with integral chokes. Part of the early integral-choke design is a steel line that runs from the passenger-side exhaust

manifold to the choke housing, also on the passenger side. Most of these engines were early L-78 396s. If you have one of these engines, attach this steel line.

After 1975, all Rochester Quadrajet engines use the integral-choke design. Connect the short U-shaped tubing from the manifold to the choke housing.

Carburetor Linkage—Rod-type carburetor linkages should have been disconnected in two places only; the throttle arm at the firewall and the transmission-kickdown rod on *some cars with automatics.* To reconnect a rod-type throttle linkage you'll need a cotter key. Make the connection through the rubber grommet. Don't lose the small plastic sleeve inside the grommet. It can pop out easily when the throttle linkage is inserted.

Reconnecting the kick-down rod involves installing a nut on a 5/16-in. hex-nut stud. This connection is made at the rear of the block and can be a minor problem. Use an ignition

Carburetor and its accessories go on as a unit. Don't overtorque carburetor bolts or nuts, 15 ft-lb is max.

Reconnect rod-type linkage at firewall arm with a new cotter key, or at carburetor with a cable-type linkage. Connect any electrical leads and double-check return spring. Check for wide-open throttle and that linkage does not bind.

wrench if you have a set, otherwise you'll probably have to use pliers to install this little nut.

Slip the cable-type linkage into the cable support mounted on the intake manifold. Pass the cable end into the clip on the carburetor throttle arm. The clip can then be placed over the arm and cable-end, and snapped in place.

Don't overlook the throttle-return spring, or you could overrev your new engine the first time you start it. This coil spring hooks to the bottom of the throttle linkage at the carburetor and runs to one of several fittings at the rear of the intake manifold. The one shown hooks to the ignition-coil bracket.

Finish the carburetor installation by connecting the pump-to-carburetor line at the carburetor inlet. *Always start this fitting* by hand or you will run the risk of stripping the threads in the carburetor. Run the fitting finger-tight. Use two wrenches, 1-in. open end at the carburetor and 5/8-in. tube nut on the fitting, to secure the fitting.

Point-System Coil—Connect the small wire from the firewall to the positive side of the coil. Marked + or −, the positive stud accepts battery power and the negative stud connects to the small wire from the distributor. It grounds the coil to the points. If you have the HEI system, you don't have to make any of these connections; the coil is in the distributor.

While you're at the back of the engine, reconnect the ground straps on both sides of the engine. Each

ground strap fits under a rear, inboard rocker-cover bolt. Also, reconnect the water-temperature- and oil-pressure-sender leads.

ACCESSORY INSTALLATION

There are huge differences between engine installations when it comes to engine accessories. It all depends on when the car was built, where it was sold and the options ordered by the customer. If you own a "stripper," with few accessories, here is your shining moment. Instead of having to deal with miles of wires and hoses or bulky and difficult-to-fit pumps or compressors, you can get on with a minimum of hassle. I will describe the procedure for installing all of the accessories fitted to big-blocks. Just skip the steps that don't apply to your engine installation.

Power Steering—On early big Chevys, two bolts thread into the front of the block and two into the side. The lower front bolt is near the bottom of the front cover and a little to the right, as viewed from the front. The top bolt goes into the block, not the head. If you can't find the bottom bolt hole, look through the hole in the pump bracket. Later versions attach the upper front bolt to the water pump, but otherwise they are the same. If the pump was removed according to my recommendations in Chapter 2, you should have nothing else to do. The lines should not have been disconnected, so the fluid reservoir should be full. Check it to make sure.

Alternator—Most early big-blocks

mount the alternator right above the power-steering pump. One bolt holds the alternator to its bracket. The alternator pivots on this bolt for belt adjustment. Run this bolt in, but don't tighten it. Part of the alternator bracket reaches around to the front of the head, where one bolt secures it. Two more bolts secure the bracket to the side of the head.

The alternator is supported at the top by a long, narrow bracket slotted for belt adjustment. Bolt the opposite end of the bracket to the corner of the head at the intersection of the head, intake manifold and block. Don't tighten this bolt just yet.

Now, for the electrical connections. One connector is secured by a stud and nut at the top, rear of the alternator. The second, and last connection, is a multi-plug connector that will plug in only one way.

Some alternators are mounted on the right side of the engine. In this configuration, the alternator pivots at the bottom on a long stud. Thread this stud into the front of the head, not the block. To install the alternator, slip it over the stud. Following the alternator, slip a triangular bracket over the stud. Secure the bracket to the water pump with two bolts and at the alternator with a nut.

The alternator is supported at the top by a triangular bracket. The pointed and slotted end of the bracket accepts the adjusting screw. The opposite end is secured by a stud and nut at the manifold and a bolt at the water pump. This upper bracket is also used on combined alternator/air-pump

Typical power-steering-pump bracket bolts to boss on water pump at front. Only earliest pumps don't mount on water pump. Exhaust-manifold bolt supports power-steering pump at rear.

Unusual long bolt/stud is for mounting alternator. Thread stud into bottom right hole in head and tighten firmly.

mountings. This bracket sometimes serves as the mount for a remote power-steering reservoir and the battery ground strap.

Air-Injection Pump—If your air-injection pump was mounted by itself, it will mount high on the passenger, or right side of the engine. This mounting uses a two-piece bracket. Instead of these two bracket pieces securing the top and bottom of the air pump, they attach more or less on each side. The right bracket runs up from the upper, right water-pump bolt to the right of the air pump. This connection is slotted for drive-belt adjustment. On the other side of the pump, the bracketry starts at the upper water-pump bolt and branches around the air pump to anchor the through bolt. Connect these two brackets to the water pump, leaving the through bolt and adjusting bolt finger-tight so you can adjust the belt later.

Alternator & Air-Injection Pump—If your alternator and air pump are mounted together on the passenger-side of the engine, bolt the three-piece bracketry to the water pump and block. If you followed the suggestion in Chapter 2, the alternator and air pump are still connected by the through bolt on the bottom of the alternator and the three brackets should still be attached to the two accessories.

Lower the assembly into the engine compartment with the alternator at the top—on large trucks, the order is reversed. The top bracket attaches to the top of the water pump. The middle bracket connects with two bolts to the bottom of the water pump. The bottom bracket attaches to the lower front of the block with two bolts. Start all bolts, then tighten them. Keep the adjusting bolts on the accessories loose for belt tightening.

Alternator connections are the same no matter how the unit is mounted. There is one stud-and-nut combination and one push-on connector. The air pump also uses the same hardware, regardless of mounting. The diverter, check valves and a lot of rubber hoses should still be connected to the air pump. Reconnect the large rubber hoses to steel lines running to the exhaust manifolds. The remaining hardware connects to the intake manifold. Follow your labeling system to reconnect air-pump plumbing.

A/C Compressor—There are two basic styles of A/C compressors. Here is how each mounts.

Early A/C compressors mount on the passenger side, fairly high on the engine. The compressor bracket is mounted to the engine with four bolts. Two are the front exhaust-manifold bolts, the other two bolts are vertically positioned at the front of the head. The compressor is adjusted similar to the alternator and has a long arm supporting the top. The engine end of this arm mounts to the head, in the same place as the alternator, except on the opposite head. The compressor end of the arm fits against a slotted portion of the compressor bracket where a bolt passes through both arm and bracket. Adjustments are made with a combination of this

Once alternator is on, slip bracket over stud and bolt to water pump. Alternator is supported at top by slotted bracket for adjusting belt tension.

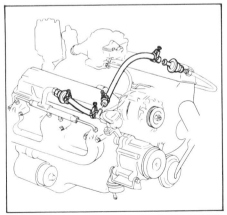

Three hoses are used with A.I.R. pump: vacuum hose from carburetor and two to the exhaust manifolds. Drawing courtesy of Chevrolet.

A/C-compressor mounts are complex, requiring a little juggling before they will line up. Front bracket must go flat against head and block, behind power-steering pump. Aluminum spacer goes between bracket and compressor at the rear.

slotted arm and bracket, and the long through bolt that holds the bottom of the compressor to the bracket.

Late A/C compressors mount on the driver side of the engine with a two-piece bracket. This bracketry is much beefier looking than the earlier style and has more bolts to secure it. The rear of the compressor is attached to a curved triangular bracket that arcs from the rear of the compressor over the rocker cover and is secured under the third and fourth intake-manifold bolts.

You'll have to remove these two bolts, fit the bracket and replace and torque these bolts 25 ft-lb. The front of the compressor bracket mounts to the side of the head with three bolts. Additionally, the front bottom of the compressor should still have a large bracket attached. Attach this with two bolts to the front of the block, behind the power-steering pump. One bolt goes right behind the power-steering pump. Installing this bolt can be tough, but it will go in! Also reattach the freon lines to the inner fender with this cross-compartment-type air conditioning.

Regardless of where an A/C compressor mounts, it is tricky to handle. The two big problems are the bulky hoses at the rear and the massive mounting brackets at both ends. If a bracket doesn't line up, use a drift punch through a mounting hole to line it up. Once you have them in, make sure the mounting bolts are tight because the compressor is heavy and vibrates a lot. Connect the electrical lead to the compressor clutch. If you don't, you'll wonder what's wrong with the air conditioning. This is the green wire just behind the compressor pulley. It has a flat, push-in connector.

Idler Pulley—Mounting the idler pulley is simple. Three bolts attach its bracket to the head. Start all the bolts, then tighten them. Pulley adjusting is done with a bolt behind the pulley, inside the bracket.

Water-Pump Pulley & Fan—The last pulley is for the water pump. The simplest way to install this pulley is to line up its bolt holes with those in the water-pump-shaft flange. Turn the flange so one of the bolt holes is straight-up. Hold the fan against the pulley with one hand, then with the other, start a bolt through the fan mounting and pulley and on into the water-pump flange. After you have two of these bolts installed you can let go of the fan. Tighten them with an open-end wrench.

Install Drive Belts—The number of belts on your engine is determined by the accessories. Few big-blocks have only one belt because almost all of the vehicles using this engine were equipped with power steering. Air conditioning requires an additional belt. If more than one belt is used, you must install them in order, from rear to front. Otherwise you won't be able to get the rear belts past the front ones.

Before you install a used belt, check it. If it's more than two-years old, replace it. If it is newer, inspect it for cracks and heavy glazing along the sides of the "V." Bend it backwards to check for cracks.

Once you have the belts on their pulleys, adjust their tension. The proper way to measure belt tension is with a belt-tension gage. But, to tell you the truth, the only place I've seen one of these tools is in a factory manual. So, if you're like me and don't have a belt-tension gage, the best way of checking belt tension is by measuring *deflection*. Check how far the belt deflects when you push on its center, between two pulleys.

All you have to do is lay a straight-edge on the back side of the belt, across two pulleys. Push the belt away from the straight-edge, midway between the pulleys. A used belt should deflect no more *or less* than 1/2 in. when under about 10 lb of force across a 14—18-in. span when it is correctly adjusted. Deflection will be slightly less with a new belt, but you should recheck belt tension *after*

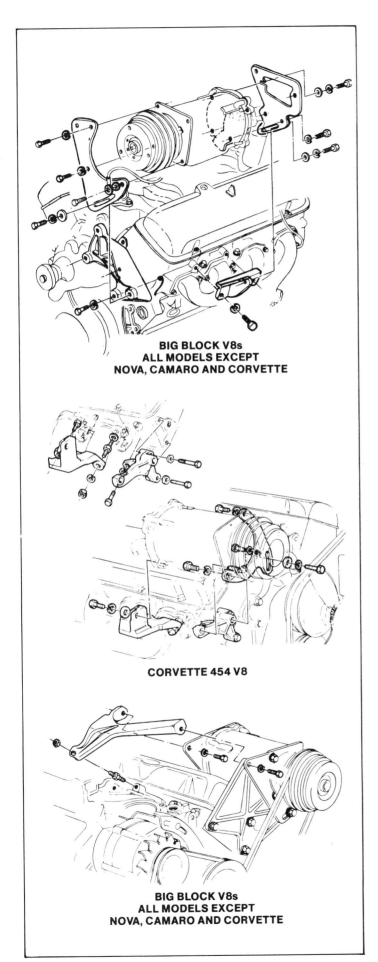

**BIG BLOCK V8s
ALL MODELS EXCEPT
NOVA, CAMARO AND CORVETTE**

CORVETTE 454 V8

**BIG BLOCK V8s
ALL MODELS EXCEPT
NOVA, CAMARO AND CORVETTE**

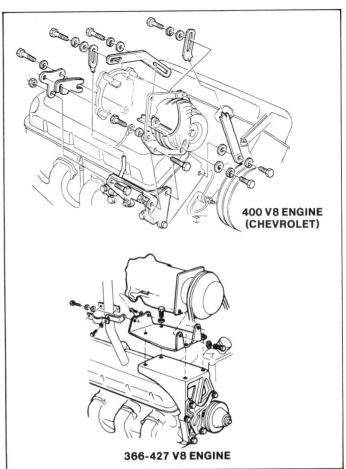

**400 V8 ENGINE
(CHEVROLET)**

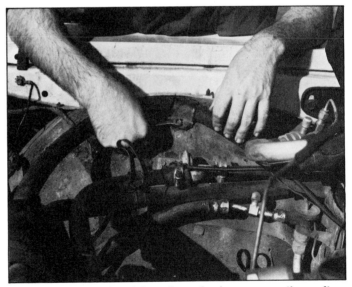

366-427 V8 ENGINE

Some more possible A/C-compressor mountings. Drawings courtesy of Chevrolet.

If A/C lines were freed from inner fenders, secure them after mounting compressor.

Install accessory-drive belts, starting with rearmost and working forward. Finish off by fitting water-pump pulley against water-pump flange.

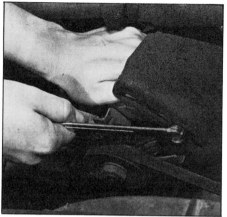

Tighten belts using adjusters on each accessory. Use deflection method to check belt tension.

Place fan shroud over fan and against front of engine *before* installing radiator.

Secure water-pump pulley and fan with four mounting bolts. Push on belt to keep pulley from turning while tightening bolts.

engine run-in. All it takes is 10 minutes of running time to break in a belt.

Why all this concern about belt tension? Simple: To start with, a too-tight belt reduces the life of the bearings that support the shafts it turns. The water pump is especially vulnerable to an overtightened belt. It will destroy the pump bearing in record time while gaining you nothing. On the other hand, if the belt is too loose, the water-pump belt will slip under load and the engine will not cool effectively. Similar problems occur with the alternator, power-steering pump and A/C compressor. Loose

belts can cause everything from a battery that continually needs charging to a squealing belt when the steering wheel is turned.

While you are adjusting belts, give the water-pump belt just a little extra slack because of the weakness of water-pump bearings—but check its tension often. The A/C-compressor belt is different. It can be considerably tighter. Keeping it so will help keep the belt from squealing.

To adjust the belt-driven accessories on a big-block Chevy, you must first loosen the through bolt running under the accessory, then the bolt passing through the slotted upper arm. Then the entire unit can be moved to loosen or tighten the belt. If an idler pulley is used, the adjustment is made at the idler. Just loosen the large nut directly behind the idler pulley. With the bolt loose, the accessory can be moved until the belt is tight, then you must hold it in place while tightening the large bolt.

Radiator & Fan Shroud—Now that all the accessories are installed on the front of the engine, the radiator and shroud can be installed. Start by placing the shroud over the fan, out of the way on the engine. If you install the radiator first, you'll have to remove it to install the shroud. The fan shroud won't fit between it and the fan. Also, make sure the shroud is right-side-up when you place it over the fan; you won't be able to turn it over without

removing the radiator.

With the shroud in place, gently lower the radiator into the engine compartment. Four bolts pass through holes in the radiator mounting plate and into clip-on captive nuts in the radiator support. Start all the bolts before tightening any of them, then run them down.

Pull the fan shroud forward to meet the radiator. Bolt the shroud to the radiator, then tighten the bolts once all four are in.

If your car has an automatic transmission, connect the two transmission cooler lines to the bottom or side radiator tank. Remove the rubber hose you placed between the two lines. Thread the tube nuts into the fittings in the tank finger-tight. Finish tightening with a 1/2-in. tube-nut wrench.

Heater Hoses—Route the heater hoses around from the passenger-side of the firewall to the front of the engine. The hoses connect at the intake manifold and the water pump. You'll be able to see where each hose should to go if you lay it across the engine. If your hoses don't seem to fall into their original routings, the right hose connects to the intake manifold and the left hose leads to the water pump. If your heater hoses come out of the firewall in a vertical plane, the top hose—heater outlet—goes to the water pump; the bottom hose goes to the intake

To protect core from damage, cover its backside with cardboard; tape it in place. Carefully lower radiator into position. Secure it with four bolts at the sides of a down-flow radiator. Cross-flow radiators are held at the top with a clamp. Remove cardboard.

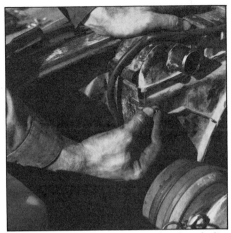

Pull shroud forward and run its four bolts in. Don't overtighten them or you'll crack the plastic.

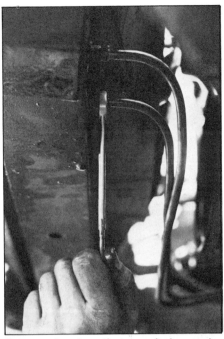

Reconnect automatic-transmission cooler lines to radiator. Snug them with a flare-nut wrench.

Connect heater hoses to intake manifold and water pump. Don't forget thermostat-bypass hose.

Alternate way of filling cooling system. Fill through top radiator hose. When coolant, or water, appears at intake manifold in thermostat opening, system is as full as you can get it for now. Install thermostat and fill radiator.

manifold. If you have not installed the water pump by-pass hose, do it now.

Like drive belts, heater or radiator hoses more than two-years old should be replaced. If you live in the hot Southwest or in a smoggy area, you should know that heat and smog have a deadly effect on rubber parts—among other things—and you can expect shorter hose and weatherstrip life in such areas. Even if your hoses are less than two-years old, check them for cracking and hardening. A cracked, brittle or mushy hose must be replaced.

Miscellaneous Hoses—Depending on the year and options, you could have anything from a nearly empty engine compartment to a bag of snakes when it comes to the small vacuum hoses so common on today's cars.

There is no way I could cover all of the different hose routings used on the big-block Chevy, so I will leave the job up to you and your photographs or notes and marking system you made during engine removal. If you marked all the hoses well, you should have no problems. If you didn't, then you are going to have to use your memory and imagination to try and figure out where they all connect.

Following is a list of the more common hoses: power-brake booster, vacuum advance, automatic-transmission modulator, various emission-control devices such as EGR, thermo-vacuum switches, vacuum motors in the air filter housing, and the climate-control vacuum motors. You really don't need to know what a lot of these hoses do as long as you marked them well so you can put them back on the correct nipples.

As you reconnect the hoses, inspect their ends for cracking. Almost all hose failures occur at the ends where the hose is stretched to go over its fitting. The hose may still look good, but it will suck air and not serve its function efficiently. Replace bad hoses as you find them. Use the old hose to gage the length of the new one.

Radiator Hoses—Next on the list are the radiator hoses. Inspect them for cracks or brittleness. Again, replace them regardless of their appearance if they are over two-years old. Clamp the hoses securely to the radiator and engine with worm-type clamps, sometimes called *aircraft clamps*. Check both hoses for clearance to the fan and accessory-drive belts.

After checking the radiator petcock to see if it closed, fill the radiator with water. It is going to take a lot of water to fill a completely dry engine. As the radiator fills, so should the block and heads. A lot of air will be trapped inside the block, so you'll hear some gurgling from the thermostat opening in the intake manifold.

Drop thermostat in place—spring side down—and mop up displaced water.

Seal thermostat housing with RTV. After hoses are connected, top off cooling system.

Keep your eye on the thermostat opening until you see water rising in the manifold. Keep filling until the water is flush with the thermostat lip. Finish filling the block through this hole. Use a funnel so you don't make a watery mess all over the engine.

It is very important to fill the block with water so the thermostat hole is completely full, even though the radiator will not accept any more water. Now, install the thermostat and its housing. Use sealer and the gasket supplied in the gasket set.

To install the thermostat, insert it into the manifold right-side up—spring into the intake manifold. If it is installed backwards, overheating will take place because the thermostat will sense the cool water in the radiator, not the hot water in the engine. Consequently, the thermostat won't open, no matter how hot the coolant gets.

Seal both sides of the gasket, affix it to the thermostat housing and place it on the block. The outlet neck points left. Torque the bolts 20 ft-lb.

Small things are easily forgotten when installing an engine. Double-check items such as the power-brake vacuum hose, water-temperature-sender unit lead and ground straps.

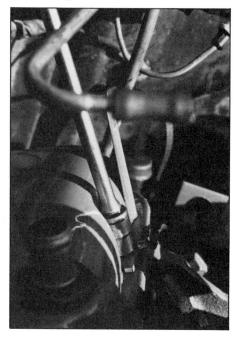

Install Battery—By now the battery should be well charged, assuming you've had it on a charger. If you haven't, take it to a gas station where it can be quick-charged. To fill the carburetor-float bowl/s, you will need a well-charged battery. The engine must have enough power to crank the engine quickly and have plenty left over to fire the plugs. Consider too that it should have sufficient reserve capacity for continued cranking if the engine won't start immediately.

After you've secured the battery in its tray, check the battery cables. They must be clean and in good condition. If they are corroded and the clamp bolts eaten away, replace them now. If

only the clamp bolt is bad, drive it out and insert a new one. Pay special attention to the possibility of the cable being eaten away at the clamp, even though it's under the insulation.

Clean the cables and battery posts with a battery cable-and-post cleaning tool. If you don't have a post cleaner it is $1 well-spent. In a pinch you can clean the posts with a wire brush and the inside of the cable ends with a rolled tube of sandpaper. Get each part shiny clean, then install the battery and cables. Coat the installed cables and posts with Vaseline or a multi-purpose, high-melting-point lube.

Time Engine—Cylinder 1 should be at 10° BTDC on its firing stroke before you can install the distributor. Remember, the crankshaft timing mark passes the timing scale twice for a complete cycle, once at the top of the intake stroke and once at the top of the power stroke.

The quick way to find the power stroke is to crank the engine with the starter. Have a friend operate the key, or hook-up a remote starter switch. Place the tip of a finger in cylinder-1 sparkplug hole, then crank the engine. Don't put your finger in past the plug threads or it could get crunched when the piston comes up.

As the piston comes up on the compression stroke you'll feel the pressure build and push your finger out of the hole. Good, the piston is coming up on the compression stroke. Now turn the crank a bump at a time until the damper timing mark is in sight. Stop cranking with the starter.

With your socket, extension and breaker bar on the damper bolt, turn the crank so the timing mark lines up with 10° BTDC. Although 10° BTDC may not be the exact timing for your engine, some ignition advance will help start the engine. You can set it to its exact specification later during its initial run-in.

Oil Pressure—Before starting your new big-block, take advantage of a neat trick with this engine—pressurizing the oiling system without starting it. This is done by turning the oil-pump drive shaft by hand before the distributor is installed. Many professional mechanics use a junked Chevy distributor for turning the oil pump. It's fast and handy.

There is an alternative: Turn the oil-

Pressurizing oil system with a speed handle and screwdriver attachment. An electric drill is faster. Regardless of what you use, turn pump clockwise.

Install a new distributor-base gasket. You'll find one in the gasket set.

Tape next to vacuum can marks cylinder 1. When correctly installed, rotor will point to tape and vacuum can will point to center of right rocker cover. Timing mark at damper should also line up with 10° BTDC.

With rotor tip about 1 in. to right of tape mark, drop distributor in place. Rotor will rotate counterclockwise as distributor gear engages cam gear. When seated, rotor and tape mark should line up. If not, pull the distributor and try again.

pump drive shaft *clockwise* with a long, 3/8-in. extension fitted with a screwdriver attachment. Use a speed handle to turn the extension or make like a fire-starting Indian and turn the extension between your palms. If you have a small-diameter extension or use a 1/4-in.-drive extension, you can chuck it into an electric drill and do it the easy way.

Actually, you don't have to turn the pump very fast for it to pump oil. Remember when you cleaned it and turned the pump by hand? You certainly moved a lot of solvent with each turn, so just turning it in your palms should be more than enough. Regardless of how you turn the pump, *make sure the extension and screwdriver attachment are tight-fitting.* If they don't fit tight, wrap any loose joints with plastic electrical tape. Check it by trying to pull the joint apart. You sure don't want to lose any tools in your engine!

If your car has an oil-pressure gage it will show when the lubrication system is pressurized. An oil-pressure warning, or *idiot*, light will indicate pressure if you have the ignition on. The light will go out as soon as pressure is indicated by the sending unit. Continue to turn the pump, however. There are a lot of galleries and lifters to fill after the oil reaches the sending unit. When your arm feels like it's going to come off at the shoulder you can stop, confident that the engine is well oiled.

Back to Timing—OK, you've got the crank correctly positioned and the oil galleries full of oil. It is now time to install the distributor.

In most engine installations you just drop the distributor in place, but with the big-block you have to position the distributor shaft and the oil-pump drive shaft so they will engage after the distributor gear engages its drive gear at the cam. Although the distributor may be correctly aligned with the cam, if the oil-pump drive-shaft slot is 90° to the distributor-shaft tang, you won't be able to install the distributor all the way. Furthermore, as the distributor gear engages the cam gear, the distributor shaft turns backwards a few degrees.

So much for the problems; let's get your distributor installed. Start by finding where the distributor rotor points in relation to its housing when it is aimed at cylinder-1 distributor-cap terminal. Do this by fitting the distributor cap to the distributor. On a points-type distributor, number-1 terminal is immediately above the point-adjustment window. If you have an HEI unit, the number-1 terminal is two plug-wire terminals to the right of the module connector when looking down on the distributor with the module connector at your left—9 o'clock. Use a felt-tip pen or strip of tape to mark the distributor housing in line with number-1 terminal.

Remove the distributor cap and turn the rotor until its metal tip points toward your mark. So far, so good. Now, take the distributor to the engine and examine the way it will fit in the intake manifold. The vacuum-advance will point to the center of the right valve cover, away from the coil. Hold the distributor in that approximate position over the engine. Note

Cylinder-1 terminal on HEI distributors is two posts clockwise away from module connector. Timing it is same as conventional models, mark cylinder-1 position with tape on distributor body. Turn rotor 1-in. to the right as you face distributor and drop the unit in place.

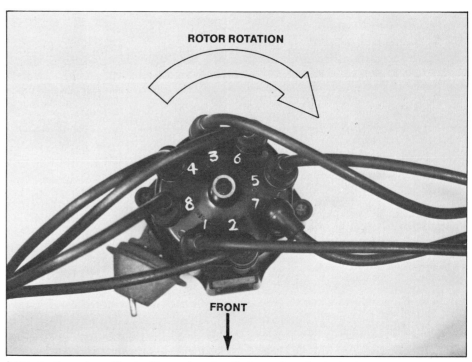

ROTOR ROTATION

FRONT

Lace up plug wires. Start with number-1 cylinder and work clockwise around distributor cap. Point-type distributor is shown. Number-1 terminal on HEI distributor is the second terminal clockwise from module connector.

the position of the oil-pump drive shaft tang in the bottom-end of the distributor gear relative to the engine. Set the distributor aside for the moment and turn your attention to the oil-pump drive shaft.

Peer into the distributor mounting hole and you should be able to see the top of the oil-pump drive shaft and its slot. If you can't, get a flashlight. Using your extension and screwdriver attachment or a long screwdriver, turn this slot so it will align with the distributor drive tang. This will allow the distributor to mate easily with the drive shaft.

Don't install the distributor just yet. You've got to make one more positioning adjustment before the distributor is ready. Because of the *helix,* or angle, of the distributor-gear teeth, the distributor will turn as you lower it in place. To compensate for this rotation, turn the distributor in the opposite direction the same amount the gears cause the rotor to turn. In the big-block Chevy, the distributor rotates clockwise. The gear-induced movement during installation is also clockwise, so you'll have to turn the rotor counterclockwise approximately 1 in. from the felt-tip mark.

Install Distributor—Put gasket on the base of the distributor. With the vacuum can pointing to the center of the right valve cover and the rotor pointing to a point 1-in. clockwise from your cylinder-1 mark, install the distributor.

Hold the distributor firmly as you lower it in place. Let the shaft and rotor turn as the gears mesh, not the housing. If the oil-pump drive shaft is lined up, the distributor should go all the way in without resistance. If it doesn't line up, check the shaft and readjust it.

When the distributor housing seats on the manifold the rotor should line up with the cylinder-1 mark. If it doesn't, note where the rotor points and remove the distributor and try again. Compensate by turning the rotor in the opposite direction, but the same distance the rotor missed its mark. When the distributor base is against the manifold and everything lines up, install the bent wire hold-down clamp and its bolt. Leave the clamp loose enough so you can turn the distributor housing.

To time the ignition, turn the distributor until the points *just open,* or, with an HEI ignition, until a tooth on

the wheel lines up with the pickup. Before you turn the distributor, check the points or teeth. One cam lobe or set of teeth should be close to the firing position—turn the distributor in that direction. The distributor should not have to be moved much, just a nudge in most cases. When you have it "timed," tighten the hold down bolt.

Distributor Cap & Wires—Complete the distributor installation by installing its cap and wires. Points-type distributors have two J-hook hold-downs, HEI caps have four. Before you turn the J-hooks, make sure the cap is positioned correctly on the distributor. Give the cap a little wiggle to make sure the cap-locating lugs are engaged with the housing. Now, turn the J-hooks.

Look at the pictures and follow the installation described in the text to install the plug wires. Use the pictures to check your work.

Lace the Wires—The firing order for *all* Chevy V-8s is 1, 8, 4, 3, 6, 5, 7, 2, and is cast into the intake manifold. The distributor turns clockwise. If it's correctly installed, cylinder-1 distributor-cap post should be slightly right of straight ahead as you look

Route plug wires through clips at rocker covers. Leads 5 and 7 should be separated to prevent misfiring.

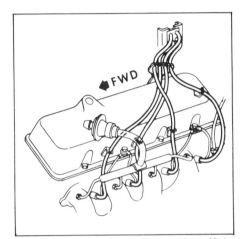

Typical left-side plug-wire routing. Note arrangement of cylinder-5 and -7 wires to gain maximum separation. Drawing courtesy of Chevrolet.

1969 L-66 air-cleaner installation is fairly simple. Later versions are complicated because of the many hoses and wires that have to be connected. Make use of your labeling system. Also fit a new base gasket and air filter to keep ring-destroying dirt out of your new engine.

down on the distributor cap. See photo on this page. That's for a points-type distributor. HEI units have the cylinder-1 post a little to the left of straight ahead.

Working clockwise, install each plug wire one at a time. Start with cylinder 1. Attach the plug-end of the wire, then route it to the number-1 post on the distributor cap. Secure the wire in the clip at the valve cover. Go around the cap in firing order to install the remaining plug wires.

All V-8s have a special arrangement for the wires going to cylinder 5 and 7 because they are next to each other both physically and in the firing order. If these plug wires are routed close and parallel to each other for any distance, *crossfiring* can occur. The current flow that fires plug 5 can *induce* a current in wire 7, causing cylinder 7 to fire 90° before it should.

You can avoid this by routing 5 and 7 wires away from each other. Start by placing them at extreme ends of their plug-wire clip.

If you are dealing with a points-type distributor, connect the small wire from the bottom of the distributor housing to the negative post on the coil. But don't install the coil-to-distributor wire yet. If you have the HEI system, leave the module plug disconnected for now.

If your Corvette has a tachometer drive, connect the tachometer cable to the drive unit and hand-tighten the knurled fitting. Don't use tools on this fitting or you will damage it.

Air Cleaner—Finish the engine installation by installing the air cleaner. Early engines need nothing more than installing the air-cleaner assembly on the carburetor with a new filter and securing it with a wing nut. Don't forget the gasket that goes between the air-cleaner base and the carburetor. This gasket must be fresh and not squashed flat or the carburetor will draw in dirty, unfiltered air.

Most engines are somewhat more complex when it comes to refitting the air cleaner. Standard smog-age air cleaners are harnessed with a breather

hose at the rear of the filter housing, hot- and fresh-air ducting, and several vacuum hoses. Make use of your labeling system or records to return these hoses to their proper fittings.

START ENGINE

The big moment has finally come. If you are like me you've got the jitters, "What did I forget to do?" Double-check to make sure you didn't overlook something.

Right now you just want to crank the engine. This will serve two purposes: it will confirm your engine can "build" oil pressure and it will fill the float bowl. Obvious things to do here are watch the oil-pressure gage or light and make sure there is fuel in the tank. Monitor the fuel situation by either watching the clear fuel filter—if you installed one—or by working the accelerator linkage while peering down the carburetor. When the float bowl fills, gasoline will squirt from the accelerator-pump nozzles. Be careful not to overdo it so you don't flood the engine.

When oil pressure and fuel are indicated, the engine is ready to start. Connect the high-tension lead to the coil, if you have a points-type

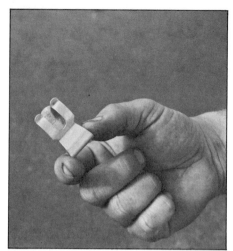

Available at most parts stores, rocker-arm clips will keep hot oil off your hands, engine, and engine compartment while setting valve lash with engine running.

Mechanical lifters mean adjusting the valves hot, or with engine running. Retorque head bolts while you're at it.

distributor; plug in an HEI-distributor connector. Hook up a tach/dwell meter, if your car doesn't have a tachometer, so you can monitor engine rpm. Finally, if an automatic transmission is used and you drained the converter, have no fewer than four-quarts of DEXRON II-type ATF on hand. You'll also need one of those long-neck funnels for adding the fluid through the dip-stick tube.

Start your engine and immediately bring it to 1500—2000 rpm. Seems like a rude awakening, but keeping rpm up is vitally important to cam and lifter life. The fast idle will help keep all internal parts well bathed in oil—a prime requirement. Letting engine rpm drop to normal idle can damage the cam and lifters, so keep engine speed up.

Right after the engine starts, add the transmission fluid. You can check its level later. Look under the car for leaks. If you find any, stop the engine and fix them. Otherwise the engine should run for about 30 minutes at fast idle before it can be slow idled. This is what it takes to break-in the cam and lifters completely.

Time Engine—With the timing light connected to the battery and cylinder-1 plug wire, loosen the distributor hold-down clamp with a 9/16-in. wrench; only enough so you can move the distributor. You want to be able to adjust the distributor and have it stay where you put it without tightening the clamp.

Turn the distributor counterclockwise to advance timing, turn it clockwise to retard it.

Be very careful when using a timing light. It is all-too-easy for timing-light leads to get caught in the fan or belts if you don't pay attention to what's going on. If that happens, the light and, even worse, your hand, will end up in the fan. Keep the timing light, its leads, your clothes and yourself away from the fan and belts. And, keep bystanders at a safe distance too.

Post Run-in Checks—After a new engine has been run for the first time, many things change. The gaskets relax, the block and heads have expanded and contracted, bolts loosen and clamps move. Go over the engine with a screwdriver and wrenches, checking every fastener in sight. Pay special attention to hose clamps. Retorque the intake-manifold and exhaust-manifold bolts. And, if other than shim-type or Permatorque composition head gaskets were used, retorque the head bolts.

Torquing the head bolts is a nasty job on most big-blocks because of the accessories that must be removed or worked around, but it is necessary. Check all fluid levels and top off as needed.

Adjust Valves—If your engine has a solid-lifter cam, you've got to readjust the valves. This means removing the air cleaner and the rocker covers, regardless of the type of head gaskets used. You may wish to drive the car for ten miles or so before adjusting the valves. But if they are really clacking, do it now and be done with it. Retorque the head bolts first, before adjusting the valves.

The valves have to be adjusted *hot*. Either make the adjustment immediately after running the engine, or take the messy approach. Adjust the valves with the engine idling. If you take the second approach, adjust the engine to slow idle. This will minimize the oil that gets flung all over you and the engine compartment. You can all but eliminate this mess through the use of *rocker-arm clips*.

Other than a socket and ratchet, you'll need a set of feeler gages. A super help when adjusting valves this way is a *go-no-go* feeler gage. But if you own a mechanical-lifter big-block, turn back to page 130 and read about it. A go-no-go gage set could be your best friend.

As for which gages you need, it depends on how much lash, or clearance, the valves must be adjusted to. The L-71 cam, used in the majority of street engines fitted with mechanical cams, including the L-71, L-72, L-78 and LS-6, require 0.024-in. clearance on the intakes and 0.028-in.

154

clearance on the exhausts. Use 0.022 and 0.024 in. for a 430-HP 427. If an aftermarket cam is used, follow the manufacturer's recommendations.

To adjust the lifters, start by warming up the engine completely. Remove one rocker cover. Intake and exhaust valves are easy to distinguish. The intakes angle in toward the carburetor at the top; the exhausts lean toward the exhaust manifold.

With a 5/8-in. socket on one end of an extension and a ratchet or breaker bar on the other, and your 0.024- or 0.028-in. feeler gage handy, start the engine. Insert the feeler gage between the rocker arm and valve tip. Loosen or tighten the adjusting nut until the proper feeler gage gives a snug fit between rocker and valve stem. Because the engine is running, you'll feel a momentary pinch on the feeler gage each time the valve opens. Go by how the gage feels when the valve is closed. This is where a slow idle helps.

Getting the proper lash while the engine is running is a little tricky, but not impossible. You'll get the most accurate adjustment if you keep the gage moving straight in and out between the valve and rocker arm. Work directly perpendicular to the centerline of the engine, not at an angle to it. When you finish adjusting one valve, go to the next one. When you finish with the intakes and exhausts in that head, replace the valve cover and do the other. Reinstall the air cleaner.

Coolant—Inspect the cooling system for leaks. If you find none, drain the water and refill the system with *coolant*. It doesn't matter where you live, antifreeze should be used. Not only is it a rust inhibitor and pump lubricant, it has a higher boiling point for summer operation. It can get hotter than pure water before it boils. In the winter, it keeps the coolant from freezing.

A 50/50 antifreeze-to-water ratio is about what is needed for the summer and winter. You can go up to a 75/25 antifreeze-to-water ratio, but the extra protection is hard to justify considering the increased cost, *unless you need really cold-weather protection.* The cooling-system capacity for the typical passenger-car-installed big-block is 21—22 quarts. Many pickup-truck engines have up to 24.5-quart systems. Heavy trucks have even larger cooling systems.

Check your owner's manual for the

Line up hood using the marks you made during removal. You'll need at least one helper to do this job.

exact specification. A 50:50 ratio for a passenger car will come to 11 quarts of antifreeze, or almost three 1-gallon jugs. This ratio provides freeze protection all the way down to −34F, probably a lot more than your engine needs. Slightly more than 7 quarts of antifreeze will give freeze protection to about 0F.

Seven quarts is one-quart short of two 1-gallon jugs. Use two complete jugs for -8F freeze protection and the boiling point will increase to over 250F if you use a 15-lb pressure cap on the radiator. Most big-blocks came with pressure caps at least this high, so you can also enjoy the security of a higher boiling point.

To install the antifreeze in your big-block, you'll need to do more than just pour it in. First warm the engine until the thermostat opens. Turn the heater on while the engine is warming up so the water in the heater will also circulate and the heater core and its hoses will fill. Position your car or truck so it is level or the front is pointed downhill so most of the water will drain from the system. Shut off the engine and drain all the water by opening the petcock at the bottom of the radiator.

Watch the water as it drains. If it is not completely colorless, fill the system again and warm the engine. Drain the coolant once more. Do this

until it comes out clean. When the system is entirely drained, it is time to install the new coolant.

Reposition your vehicle so it now points uphill for filling the cooling system. It's OK to run the engine for a short while without coolant. Close the petcock and partially fill the system with water. Then pour in the antifreeze and top it off with water. Start the engine and turn on the heater. While keeping an eye on the water-temperature gage, run the engine until the thermostat opens. If you get an overheating condition, stop the engine. *After it cools,* fill the block through the thermostat opening.

Sometimes it is hard to get a big-block to "burp" itself. The thermostat will open, but bubbles will rise occasionally in the radiator and, if you measured the water poured into the engine, you would know the cooling system was not full. If your big-block seems to have air trapped in it, take it out for a short drive, or at least fast-idle it. The higher water-pump speed will help purge trapped air from water-jacket pockets. And, the higher the filler neck, the better. I've even seen Corvettes with their right side raised with a floor jack so the surge tank would be higher. This allows trapped air to escape more easily from the cooling system. You might try it if you have a 'Vette.

Recheck all hoses and exposed bolts for tightness. Check all fluid levels and look for leaks. As you accumulate miles on the new engine, it is important to watch the fluid levels and top off as necessary. It is a bad idea to plan a long trip using the rebuilt engine as there's always going to be some little thing that needs attention. If you are near home it's no problem, but out on the road little problems have a way of turning into big problems.

Drive your vehicle normally, but avoid full-throttle acceleration and steady-speed cruising. Go easy and vary your speed. It is normal for a new engine to use some oil, so don't worry if yours does. When an engine using moly rings has been rebuilt correctly it usually won't take more than 200–300 miles for it to break-in. If by 500 miles your engine still hasn't broken-in, something is wrong.

Those with chrome rings take longer to seat, so oil control may be poor well past 1000 miles.

After putting 500 miles on your new engine take it to a tuneup shop and have it "gone through." This will ensure that the engine runs at peak efficiency, yielding the best balance between maximum power, fuel economy and low emissions. Don't take the car to the tuneup shop immediately after the rebuild. The final internal dimensions will not exist and the tuneup will "go off" after a very short while. Have the engine tuned only after it is fully broken-in.

When you're satisfied the cooling system is full of liquid, cap the radiator and install the hood.

Hood—You'll need someone to help with installing the hood. Hold the hood in place and run all the bolts in finger-tight. Then position the hood to the reference marks you made or the holes you drilled through the hinges and hood inner panel. Tighten the hood bolts and close the hood *slowly* to check its fit. If the hood is out of place, loosen the bolts slightly and bump it into place. Retighten the bolts and recheck its fit.

First Drive—There are several items to keep in mind when taking your big-block out for its first drive. First, make an around-the-car inspection for tools and other objects that may have found their way near or under the car. Check the trunk and under-hood areas for tools or parts. You sure don't want to distribute your tools on the road, so make sure they are all safely back in your toolbox. For your first drive don't take the car on an errand where your attention may be focused on something other than the engine. Instead, go for a ten-minute drive just to feel out the engine, then return for an underhood inspection.

OFFICIAL SHOP MANUALS

For additional information concerning you car or truck, you should obtain the official Chevrolet or GMC shop manual. These manuals are available from Helm. To order, or for additional information, contact Helm, Inc., P.O. Box 07130, Detroit, MI 48207. Telephone: (313) 865-5000. Make sure you include the model and year of your vehicle when ordering.

9 Tuneup

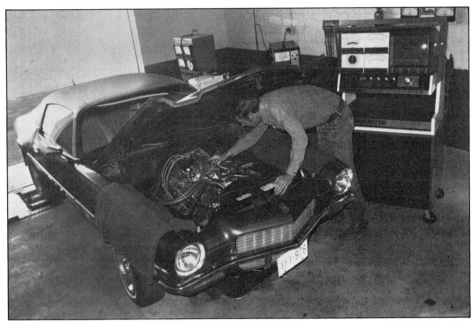

After your engine is broken in—300—500 miles—take it to a tuneup specialist. Have him tune your new engine for maximum performance and minimum emissions. Photo by David Vizard.

Electronic engine analyzers are capable of testing a number of engine functions. Twenty minutes on one of these is worth two weeks of shade-tree knuckle busting. Stack of gages at left are used for testing plug wires, condenser, distributor cap, perform a power-balance test and leak-down test, set dwell, check points resistance, test HEI functions and much more.

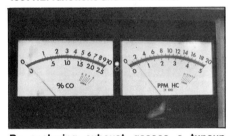

By analyzing exhaust gasses a tuneup mechanic can learn a lot about air/fuel ratio to obtain overall efficiency. Exhaust-gas analysis plays a big part in modern engine tuning.

Now that you have invested a considerable amount of time and money into your big-block Chevy, it makes sense to do everything to keep your new engine running at peak efficiency. In short, have your Chevy tuned up.

Before you run for the timing light and dwell-meter, I would like to stress the value of professional tuneup specialists. First of all, you've already set the timing on your new engine, it has fresh plugs. Maybe you rebuilt the distributor and carburetor as well. All internal pieces are new or reconditioned and were correctly installed, so why get the engine tuned? Right now there is no need, but in several hundred miles there will be. Allow the engine to break-in, which will occur within 500 miles, then take it to a tuning specialist. The specialist has the *tools* and *skills* to adjust your big-block to maximum operating efficiency.

The best tuneups are done on a *chassis dynamometer,* a device that has wheels in the floor that are driven by the drive wheels of your car or truck. This simulates very accurately the real-world conditions while allowing the mechanic to monitor and adjust engine performance with an oscillo-scope while your vehicle is "driving down the road."

In addition, a chassis dyno is capable of monitoring drive-wheel horsepower, which will tell the whole story about any engine modifications you might have made. As a cautionary note you should know that although you can get power gains by tuning your engine for maximum horsepower at full-open throttle, you should not do this because it will hurt engine efficiency at other, more common, operating conditions.

Most tuneup shops do not have a chassis dyno, but are equipped with an oscilloscope, or *electronic engine analyzer* as it is sometimes called. These shops come a *very* close second to chassis dyno shops in results. You probably couldn't feel the difference between the two different tuneups.

Last in line, of course, is you and your timing light. Unfortunately you just can't tune a car, using home equipment, to the same fine edge as a pro with his sophisticated equipment.

What this all means to you is your big-block needs a little time to break in, at which point it should go to a pro for a tuneup. He'll find and correct any problems that may have been present—and usually there are a couple in a fresh rebuild—while ensuring that the engine performs at least as well as it did when it came out of the factory. You will enjoy maximum power and economy for a moderate cost. Prices vary from $35 to $75 for tuneups and, because most of your parts are new and will not need replacement, you should end up at the low end of this range.

In addition to the performance and economy advantages of tuning, there is the emissions consideration. Properly tuned engines produce less pollutants, so you will be doing your part for cleaner air by keeping your big Chevy in top tune. If you are planning on selling your big-block-powered vehicle, you may have to certify it in a state-run emissions program before the title can be transferred. You may already be doing this for yearly registration. If you keep your engine freshly tuned it will pass its exams more easily.

Index

A

A/C
 install 145-148
 refrigerant 18, 19
 remove 18, 19
Air cleaner 18, 153
Air pump (A.I.R.) 20, 145
Align boring 73
Alternator 20, 22, 23, 144, 145
Anti-seize compound 109

B

Balance
 external 33, 36, 37
 heavy metal 35
 internal 33, 36, 37
Ball gage 98
Battery 15, 150
Belts 146, 148
Blowby 4, 13
Bore wear 67
 taper 67-69
Boring 33, 34
Bottom end 6

C

Camshaft 80-82
 gear drive 82
 inspection 81, 82
 install 112, 113
 plug 111
 remove 60, 61
Camshaft bearing 62
 chamfer 110
 grooved 34, 110, 111
 install 110, 111
Camshaft-lobe wear 12, 13
Carbon deposits 5, 6, 96
Carburetor
 choke 51, 143, 144
 install 143
 remove 51, 52
Casting number 30-32, 36, 40
Chain
 lifting 23
Chassis dynomometer 157
Clutch 49, 137-140
Compression tester 10
Coolant 16, 155
Cooling lines, tranmission
 install 149
 remove 17
Connecting rod
 bolts 88, 89
 journal protectors 59, 121
 inspection 87
 identification 37
 out-of-round 88
 number 57
 recondition 88
 size 88
Compression height 38, 39
Core plugs 50, 111
Clutch linkage 22
Crankshaft 75, 80
 edge riding 78
 end play 117
 forged steel 36
 identification 35-37
 install 113, 115, 116
 journal clearance 114

kits 77
 nodular iron 36
 out-of-round 75, 76
 pilot bearing 79
 Plastigage 116
 polish 78
 rear seal 113, 116, 117
 remove 59-61
 runout 77
 swapping 35, 36
 taper 75-77
Crankshaft damper
 balance 35
 inspection 125
 install 125
 remove 55, 56
 repair sleeve 125
Crankshaft pulleys
 install 135
Cylinder block
 cleaning 64-66
 decking 71, 72
 identification 31-35
Cylinder head
 cleaning 96, 128
 disassembly 94-96
 identification 40-44
 install 128, 129
 milling 97
 remove 53-55
 warpage 97

D

Deck height 31
Detonation 5
Diagnosis 6-9
Dial indicator 12, 77, 117
Dial bore gage 67-69
Distributor
 cap and wires 152, 153
 gasket 151
 install 151, 152
 remove 51, 52
Dipstick
 install 133
 remove 53
Drive shaft, oil pump 57, 91, 125, 126, 151, 152
 connector 57, 126

E

End clearance 91
Engine
 install 140-156
 mount 22, 134, 135, 142
 noise 6-8
 remove 14-24
 stand 50
 weight 3
Engine analyzer 157
Engine hoist 14, 23
Engine production code 27, 30
Engine suffix 27-30
Exhaust manifold 20, 21, 142
 identification 46, 47
 install 132, 133
 remove 52, 53
 seals 140
Exhaust pipe 20, 21, 142

F

Fan 18, 146

Fan shroud 17, 148, 149
Firing order 152
Flare-nut wrench 17, 51, 148
Flexplate 21, 24, 35, 49, 139, 140
Flywheel 49, 137-140
Front cover
 install 124, 125
 remove 56
Fuel filter
 install 134, 135, 143
Fuel pump 20
 eccentric 134
 install 134
 noise 8
 pushrod 53, 134
 remove 53

G

Gallery plugs 61, 111, 112
 vented 112
Go-no-go gage 154
Glaze breaking 71
Graph
 Taper vs ring end-gap 70
Ground straps 23, 150

H

Head gasket
 blown 5
 doubling 93
 install 128
 sealing 128
Heater hose
 install 148, 149
High-tension lead 9
Heat-riser valve
 install 142, 143
 noise 8
 remove 21
Hot tank 62, 65, 74
Hood
 install 156
 remove 16

I

Ignition
 coil 50
 connect 144
 disable 9
 HEI 9
 points-type 9
 timing 150, 151, 154
 wires 9, 10, 52, 152, 153
Intake manifold
 cleaning 107
 identification 44-46
 install 131, 132, 135
 milling 97
 remove 53

K

Knock 5
Knurl
 pistons 71
 valve guides 98, 99

L

Lapping
 oil pump 91
 valves 103
Leak-down test 10, 11
Lifters 80-83
 cleaning 82, 83

diagnosis 7, 8
 inspection 82, 83
 install 129
 remove 60, 61
Lifting plates 23

M

Magnaflux 74
Major dimension 76
Main bearing
 bores 72, 73
 crush 87
 diagnosis 7, 63
 grooved 113
 install 114, 115
 remove 59, 60
 size 113
Mark IV 3, 26
Metric/Customary-Unit
 Equivalents 158
Molybdenum-disulfide 109

N

Notching 64
Noise
 engine 6-8

O

Oil 108
Oil clearance 76
Oil consumption 4
Oil filter
 adapter 62, 63
 cartridge vs spin-on 62, 63
 install 140
Oil pan
 install 126-128
 remove 57
Oil-pressure sender
 install 134, 135
 remove 21, 52, 53
Oil pump
 drive shaft 57, 91, 125, 126, 151, 152
 end clearance 91
 install 126
 pressure-relief valve 92
 remove 57
 teardown 91
Out-of-round 75

P

Part number 30
PCV system 4, 5, 13
Pilot bearing 79, 80
Pinging 5
Piston
 clearance 70, 71
 compression height 38, 39
 dome 38, 85
 identification 38
 inspection 83-86
 knurling 71
 noise 7
 ring grooves 86, 87
 slap 6, 7
Piston pin 37
Piston ring 4, 5
 compressor 109
 dot 119
 end gap 117, 118
 expander 109, 110

gap spacing 120
install 118-120
twist 119
types 73
Piston and rod
assembly 89, 90
disassembly 89, 90
install 120-123
remove 58, 59
Plastigage 109, 114
Power steering
install 144
remove 19
Preignition 5
Protect-O-Plate 26, 27
Pressure plate
see clutch
Puller 55, 56
Pushrods
identification 44
inspection 130
installation 129
remove 54

R
Radiator
install 148
remove 17, 18
Release bearing 139
Ridge reaming 57
Ring expander 83, 119, 120
Rocker arm
ball 54
install 129
remove 54
stud 96
Rod bearing
crush 87
diagnosis 7, 63, 88
Plastigage 114
remove 59

RPO 25, 27, 48
Runout 77

S
Sealer 108
Sleeve 35, 68
Slide hammer 79
Small-hole gage 98
Sparkplug
install 133
reading 8, 9
style 43
Spotcheck 74
Starter motor 21, 22, 142

T
Tables
Block-Casting Numbers 32
Camshaft-Forging Numbers 31
Camshaft-Lobe Lift 13
Crankshaft-Bearing Journals 36
Crankshaft-Casting Numbers 36
Cylinder-Head Casting
Numbers 40
Engine-Series Number &
Suffix 28, 29
Head Games 43
Passenger-Car RPO
Descriptions 48
Protect-O-Plate 26
RPO's 27
Taper 70
Valve-Spring Specifications 104
Valve Stem-To-Guide
Clearance 98
Vehicle Identification Number 26
Tall block 31
Taper 75
Tappet 6
Tests
combustion leak 11

compression 10
cranking vacuum 9
leak down 10, 11
power balance 10
Thermostat
bypass 49, 149
install 150
remove 50
test 135
Throttle linkage 18, 51, 143, 144
Throttle-return spring 144
Thrust side 7
Timing chain
inspect 56
install 117, 123, 124
remove 56, 57
types 33, 90, 91
Timing light 6, 154, 157
Torque converter 21, 24, 136, 137
Torque wrench 109
Transmission front seal 136, 137
Tufftriding 77
Tuneup 5, 157

V
Vacuum
ported 9
Vacuum gage 9
Vacuum hoses 15, 18, 149
Valve
adjust 130, 131, 154, 155
burned 5, 6, 100-102
grind 101, 102
lift 12, 13
oversize 99
remove 95
rotators 96
shrouding 6, 41
size 42, 43

stem wear 101
stuck 11, 12
Valve cover
install 132
remove 54
Valve guide 4, 43, 97
insert 99, 100
knurl 98, 99
measure 98
wear 98
Valve seat
recondition 102, 103
Valve spring
broken 12
compressor 94, 95, 107
dual 95, 96, 107
free heigh 104
inspect 103-106
install 106, 107
installed height 106
keeper, 95, 107
load, installed/open 104
rate 103, 104
retainer 107
specifications 104
test 105
VIN 25, 26

W
Water injection 39, 93
Water pump
install 131
remove 49, 50
Water-temperature sender
install 135
location 44
remove 19, 54
Wires, ignition 9, 10, 52, 152, 153
Wrist pin 6

METRIC/CUSTOMARY-UNIT EQUIVALENTS

Multiply:	by:	to get:	Multiply:	by:	to get:

LINEAR

inches	X 25.4	= millimeters(mm)	X 0.03937	= inches
feet	X 0.3048	= meters (m)	X 3.281	= feet
miles	X 1.6093	= kilometers (km)	X 0.6214	= miles

AREA

$inches^2$	X 645.16	= $millimeters^2(mm^2)$	X 0.00155	= $inches^2$
$feet^2$	X 0.0929	= $meters^2(m^2)$	X 10.764	= $feet^2$

VOLUME

$inches^3$	X 16387	= $millimeters^3(mm^3)$	X 0.000061	= $inches^3$
$inches^3$	X 0.01639	= liters (l)	X 61.024	= $inches^3$
quarts	X 0.94635	= liters (l)	X 1.0567	= quarts
gallons	X 3.7854	= liters (l)	X 0.2642	= gallons
$feet^3$	X 28.317	= liters (l)	X 0.03531	= $feet^3$
$feet^3$	X 0.02832	= $meters^3(m^3)$	X 35.315	= $feet^3$

MASS

pounds (av)	X 0.4536	= kilograms (kg)	X 2.2046	= pounds (av)

FORCE

pounds—f(av)	X 4.448	= newtons (N)	X 0.2248	= pounds—f(av)
kilograms—f	X 9.807	= newtons (N)	X 0.10197	= kilograms—f

TEMPERATURE

Degrees Celsius (C) = 0.556 (F - 32)   Degrees Fahrenheit (F) = (1.8C) + 32

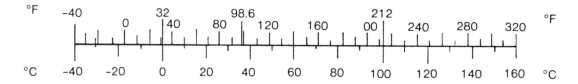

PRESSURE OR STRESS

inches Hg (60F)	X 3.377	= kilopascals (kPa)	X 0.2961	= inches Hg
pounds/sq in.	X 6.895	= kilopascals (kPa)	X 0.145	= pounds/sq in

POWER

horsepower	X 0.746	= kilowatts (kW)	X 1.34	= horsepower

TORQUE

pound-inches	X 0.11298	= newton-meters (N-m)	X 8.851	= pound-inches
pound-feet	X 1.3558	= newton-meters (N-m)	X 0.7376	= pound-feet
pound-inches	X 0.0115	= kilogram-meters (Kg-M)	X 87	= pound-feet
pound-feet	X 0.138	= kilogram-meters (Kg-M)	X 7.25	= pound-feet

VELOCITY

miles/hour	X 1.6093	= kilometers/hour(km/h)	X 0.6214	= miles/hour
kilometers/hr	X 0.27778	= meters/sec (m/s)	X 3.600	= kilometers/hr

COMMON METRIC PREFIXES

mega	(M)	= 1,000,000	or 10^6	centi	(c)	= 0.01	or 10^{-2}
kilo	(k)	= 1,000	or 10^3	milli	(m)	= 0.001	or 10^{-3}
hecto	(h)	= 100	or 10^2	micro	(μ)	= 0.000,001	or 10^{-6}

Conversion chart courtesy of Ford Motor Company.